THE BEHAVIOURAL BIOLOGY OF DOGS

The Behavioural Biology of Dogs

Edited by

Per Jensen
IFM Biology
Linköping University
Sweden

CABI is a trading name of CAB International

CABI Head Office
Nosworthy Way
Wallingford
Oxfordshire OX10 8DE
UK

Tel: +44 (0)1491 832111
Fax: +44 (0)1491 833508
E-mail: cabi@cabi.org
Website: www.cabi.org

CABI North American Office
875 Massachusetts Avenue
7th Floor
Cambridge, MA 02139
USA

Tel: +1 617 395 4056
Fax: +1 617 354 6875
E-mail: cabi-nao@cabi.org

© CAB International 2007. All rights reserved. No part of this publication may be reproduced in any form or by any means, electronically, mechanically, by photocopying, recording or otherwise, without the prior permission of the copyright owners.

A catalogue record for this book is available from the British Library, London, UK.

A catalogue record for this book is available from the Library of Congress, Washington, DC.

ISBN-10: 1 84593 1874
ISBN-13: 978 1 84593 1872

Typeset by MRM Graphics Ltd, Winslow, Bucks.
Printed and bound in the UK by Cromwell Press, Trowbridge.

Contents

Contributors	vii
Preface	ix
I: The Dog in Its Zoological Context Editor's Introduction	1
1. Evolutionary History of Canids *X. Wang and R.H. Tedford*	3
2. Domestication of Dogs *P. Savolainen*	21
3. Origin of Dog Breed Diversity *C. Vilà and J.A. Leonard*	38
II: Biology and Behaviour of Dogs Editor's Introduction	59
4. Mechanisms and Function in Dog Behaviour *P. Jensen*	61
5. Behaviour Genetics in Canids *E. Jazin*	76

6. Sensory Physiology and Dog Behaviour — 91
H. Bubna-Littitz

7. Social Behaviour of Dogs and Related Canids — 105
D.U. Feddersen-Petersen

8. Learning in Dogs — 120
P. Reid

III: The Dog in Its Niche: Among Humans — 145
Editor's Introduction

9. Behaviour and Social Ecology of Free-ranging Dogs — 147
L. Boitani, P. Ciucci and A. Ortolani

10. Evolutionary Aspects on Breeding of Working Dogs — 166
R. Beilharz

11. Individual Differences in Behaviour – Dog Personality — 182
K. Svartberg

12. Human–Animal Interactions and Social Cognition in Dogs — 207
Á. Miklósi

IV: Behavioural Problems of Dogs — 223
Editor's Introduction

13. Behavioural Disorders of Dogs — 225
R.A. Mugford

14. Behaviour and Disease in Dogs — 243
Å. Hedhammar and K. Hultin-Jäderlund

Index — 263

Contributors

Beilharz, R., *School of Agriculture and Food Systems, Faculty of Land and Food Resources, The University of Melbourne, Melbourne, Victoria 3010, Australia*
Boitani, L., *Department of Animal and Human Biology, University of Rome, Viale Università 32, 00185 Roma, Italy*
Bubna-Littitz, H., *Department for Natural Sciences, Institute for Physiology, Veterinary University of Vienna, Veterinärplatz 1, A-1030 Vienna, Austria*
Ciucci, P., *Department of Animal and Human Biology, University of Rome, Viale Università 32, 00185 Roma, Italy*
Feddersen-Petersen, D.U., *Department of Zoology, University of Kiel, Olshausenstr. 40, D-24118 Kiel, Germany*
Hedhammar, Å., *Department of Small Animal Clinical Sciences, Swedish University of Agricultural Sciences, Box 7037, SE-750 07 Uppsala, Sweden*
Hultin-Jäderlund, K., *Department of Small Animal Clinical Sciences, Norwegian College of Veterinary Medicine, PO Box 8146 Dep., N-0033 Oslo, Norway*
Jazin, E., *Department of Evolution, Genomics and Systematics, Uppsala University, Norbyvagen 18D, SE-752 36 Uppsala, Sweden*
Jensen, P., *IFM Biology, Linköping University, SE-581 83 Linköping, Sweden*
Leonard, J.A., *Department of Evolutionary Biology, Uppsala University, Norbyvägen 18D, SE-752 36 Uppsala, Sweden*
Miklósi, Á., *Department of Ethology, Eötvös Loránd University, Pázmány sétány 1/c 1117, Budapest, Hungary*
Mugford, R.A., *The Animal Behaviour Centre, PO Box 23, Chertsey, Surrey KT16 9NL, UK*
Ortolani, A., *Department of Animal and Human Biology, University of Rome, Viale Università 32, 00185 Roma, Italy*
Reid, P., *ASPCA Animal Behavior Center, 424 East 92nd Street, New York, NY 10128, USA*

Savolainen, P., *Department of Biotechnology, Albanova University Centre, KTH-Royal Institute of Technology, SE-10691 Stockholm, Sweden*

Svartberg, K., *Department of Anatomy and Physiology, Swedish University of Agricultural Sciences, PO Box 7011, SE-750 07 Uppsala, Sweden*

Tedford, R.H., *Department of Paleontology, American Museum of Natural History, Central Park West at 79th Street, New York, NY 10024-5192, USA*

Vilà, C., *Department of Evolutionary Biology, Uppsala University, Norbyvägen 18D, SE-752 36 Uppsala, Sweden*

Wang, X., *Department of Vertebrate Paleontology, Natural History Museum of Los Angeles County, 900 Exposition Blvd., Los Angeles, CA 90007, USA and Department of Paleontology, American Museum of Natural History, Central Park West at 79th Street, New York, NY 10024-5192, USA*

Preface

Dogs are our oldest domesticated animals, our friends and companions, and the most widespread species of all animals under human care. There are few things for which dogs have not been utilized during our joint history: they have been used as hunting aids, for pulling and carrying, as providers of meat, fur and other products, for guarding and watching, and as laboratory test animals, just to mention some examples. In large parts of the world, dogs are increasingly popular as pets, friends and family members, and at the same time the importance of dogs as working dogs, for example with police and rescue forces, has probably never been higher.

At the same time, biologists have started to take greater interest in the biology and behaviour of dogs, for a number of reasons. First, since dogs are so closely connected to human evolution and history, understanding how, where and when dogs have developed and spread helps us understand our own background. Second, the long coexistence with humans has led the dog to develop specific adaptations facilitating life with us, and the dog therefore provides an excellent model for studying how behaviour and cognition have evolved. Third, the increasing importance of dogs as pets and working animals calls for a deeper biological knowledge of how these animals actually work – such information can help us not only to train and shape dogs for specific tasks, but also to prevent and cure various behavioural disorders which may cause owners and animals large problems.

The present book is an attempt to provide an up-to-date description of the behavioural biology of dogs, written by experts in different areas of this large field. The target audience consists of students of animal behaviour or veterinary medicine at advanced levels – the book is not intended as an introductory text to dog behaviour in general. It is also the hope of the authors that interested dog owners outside academia may find usable parts in the book. There is no doubt that certain chapters will require closer acquaintance with various aspects of biology than the average dog owner is likely to possess, but we also believe that most chapters contain aspects that are readily accessible.

The book is split into four different parts, each concerned with a specific aspect of the behavioural biology of dogs. The first part (Chapters 1–3) is devoted mainly to the evolution and development of the dog. Although not primarily concerned with behaviour, these aspects form the basis for understanding how behaviour has developed and for placing the dog in its relevant biological context. The second part (Chapters 4–8) deals with basic aspects of animal behaviour with particular emphasis on dogs. The third part (Chapters 9–12) places the modern dog in its present ecological framework: in the niche of human coexistence. Here we give a broad overview of the behavioural aspects of living close to humans. In the last part of the book (Chapters 13 and 14), the emphasis is on behavioural problems, their prevention and cure.

All of the contributors to this book have considerable research experience in their areas, and it is hoped that this will guarantee that the text is relevant, up-to-date and central to the subject.

Per Jensen
Linköping, March 2006

I The Dog in Its Zoological Context

Editor's Introduction

In this first part of the book, the dog is placed in context amongst its zoological relatives, and in relation to its domestication history. The first chapter outlines the modern view on the zoological systematics of carnivores and canids in particular. Here, the reader will find an exciting account of fossil and present traits which allows the dog to be placed within the greater picture of closely related canids.

Of course, the dog is a domesticated species, and its domestication history has been subject to intense research during the last decade or so. Here, modern molecular genetics offers tools which have allowed biologists to give their pictures of how, where and when domestication started, complementing the traditional picture offered mainly by archaeologists. Some aspects of this new picture are truly stunning and require that we revise large parts of our traditional views on how domestication began.

Given the novelty of the molecular research on the ancestry of dogs, it should not be a surprise that different scientific groups arrive at somewhat different conclusions. This is partly explained by variations in methods, and the only way to resolve some of the disagreements is to continue to do more and improved research. As this book is written, there is therefore only limited consensus among biologists concerning, for example, the time when domestication started. As editor, one has to make a decision – one can choose not to cover the new research at all, or to only present the picture oneself believes in, or to present divergent pictures and allow the reader to decide. I have chosen the last strategy, and therefore, the stories presented in Chapters 2 and 3 differ on some important points – these differences will hopefully disappear in the light of future research.

Evolutionary History of Canids

Xiaoming Wang and Richard H. Tedford

Introduction

Members of the dog family (Canidae) are an early lineage of carnivorans (order Carnivora). The canids were the first to branch off the caniform carnivorans, dog-like predators that, in addition to canids, also include the bear family (Ursidae), the raccoon family (Procyonidae), the weasel family (Mustelidae), as well as the aquatic seals, sea lions and walruses (Pinnipedia). Living canids are some of the most successful predators, occupying all continents except Antarctica, and reign supreme as top predators in parts of northern North America and Eurasia. It is thus quite remarkable that domestic dogs, known for their loyalty to human masters, came from a dominant predator in the form of the grey wolf.

The history of domestic dogs occupies a tiny fraction of the long family history, and represents a mere twig in a large family tree of more than 36 species of wild canids living today (Wang *et al.*, 2004a, b). Despite their impressive variety, all dogs came from a single species of wolves in the latest Pleistocene during the last Ice Age. Earliest fossil evidence of domestic dogs in archaeological sites dates around 12,000–14,000 years ago in western Eurasia, whereas genetic evidence suggests an East Asian origin around 15,000 years ago or possibly as old as 100,000 or more years ago (e.g. Leonard *et al.*, 2002; Pennisi, 2002; Savolainen *et al.*, 2002). Often treated as a subspecies of the grey wolf, *Canis lupus familiaris* (Linnaeus, 1758), domestic dogs bear numerous resemblances to their wild wolf ancestors both morphologically and behaviourally.

The evolutionary history of canids is a history of successive radiations repeatedly occupying a broad spectrum of niches ranging from large, pursuit predators to small omnivores, or even to herbivores. Three such radiations were first

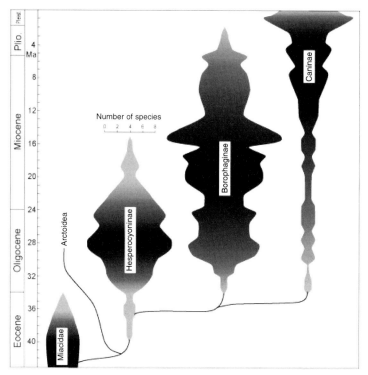

Fig. 1.1. Range of distribution, diversity through time and relationships among the three subfamilies of Canidae. The horizontal width of each subfamily range roughly reflects the total number of species at a given time, as indicated by scale bar above Hesperocyoninae. That for the family Miacidae is not meant to be strictly proportional to the species diversity.

recognized by Tedford (1978), each represented by a distinct subfamily (Fig. 1.1). Two archaic subfamilies, Hesperocyoninae and Borophaginae, thrived in the middle to late Cenozoic from about 40 to 2 million years ago (Ma) (Wang, 1994; Wang et al., 1999). Living canids all belong to the final radiation, subfamily Caninae, which had almost an equally long history as the hesperocyonines and borophagines but achieved their present diversity only in the last few million years (Tedford et al., 1995).

Canids originated more than 40 Ma in the late Eocene of North America from a group of archaic carnivorans, the Miacidae (Wang and Tedford, 1994, 1996). They were confined to the North American continent during much of their early history, playing a wide range of predatory roles that encompassed those of the living canids, procyonids (raccoons), hyaenids (hyenas) and possibly felids (cats). By the latest Miocene (about 7–8 Ma), members of the subfamily Caninae were finally able to cross the Bering Strait to reach Europe (Crusafont-Pairó, 1950), commencing an explosive radiation and giving rise to the modern canids of the Old World. At the formation of the Isthmus of Panama about 3 Ma, canids arrived in South America and quickly established themselves as one of the most diverse groups of predators on the continent (Berta, 1987, 1988). With the aid of humans, *Canis lupus dingo* was transported to Australia late in the Holocene. Since that time, canids have become truly worldwide predators, unsurpassed in distribution by any other group of carnivorans.

Here, in the context of this volume on domestic dogs, we place more emphasis on the subfamily Caninae, to place the origin of the wolves in its proper historical context. The issues of dog domestication will be treated separately (Chapter 2 in this volume). Our perspectives are mostly palaeontological and morphological, although we will point out controversies from molecular studies. We do not attempt to cite all of the references in canid palaeontology and systematics, most of which have been summarized in papers that are cited.

What is a Canid?

Canids are members of the order Carnivora due to their common possession of a pair of carnassial teeth. The carnassials are formed by the upper fourth premolar and lower first molar, which have long, sharp shearing blades and function as a pair of scissors for slicing muscles and tendons of the prey. Within the Carnivora, canids fall in the suborder Caniformia, or dog-like forms. The Caniformia are divided into two major groups that have a sister relationship: superfamily Cynoidea, which includes Canidae, and superfamily Arctoidea, which includes the Ursidae, Ailuridae, Procyonidae and Mustelidae, as well as the aquatic Pinnipedia and the extinct Amphicyonidae.

The family Canidae is a group of carnivorans that originated from a common ancestor more than 40 Ma. Through such a common evolutionary ancestry all members of the Canidae share a few morphological features (shared derived characters) that are passed to all their descendants, although some of these features have been modified subsequently in different ways. All of these derived characters can be observed in the fossil record, and thus are capable of being verified time and again throughout their evolutionary history.

As a cohesive group of carnivorans, living canids are easily distinguished from other carnivoran families. Morphologically there is little difficulty in recognizing living canids with their relatively uniform and unspecialized dentitions. However, the canids as exemplified by the living forms are narrowly defined. Only a small fraction of a once diverse group has survived to the present day (Fig. 1.1). Canids in the past had departed from this conservative pattern sufficiently that palaeontologists had misjudged some canids as procyonids. Similarly the extinct bear-dog family Amphicyonidae, which belongs to the Arctoidea, had in the past been placed within the Canidae, because of its unspecialized dentition.

How do we know a canid when we see one? A key region of the anatomy used to define canids is the middle ear region, an area in the back of the skull that displays a rich variety of morphological patterns (Hunt, 1974). In particular, the way the middle ear bullar chamber (a rounded bony housing that protects the

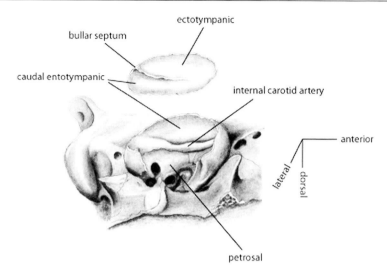

Fig. 1.2. Ventrolateral view of basicranial morphology of a primitive canid, *Hesperocyon gregarius*, showing bullar composition and position of the internal carotid artery (see text for explanations). The ventral floor of the bulla is dissected away (isolated oval piece on top) to reveal the middle ear structures inside the bulla and the internal septum. Modified from Wang and Tedford (1994, Fig. 1).

inner ear) ossifies is of considerable importance in recognizing relationships among different families of carnivorans. Canids are characterized by an inflated entotympanic bulla that is divided by a partial septum along the suture between entotympanic and ectotympanic bones that form the bulla enclosing the floor of the middle ear space (Fig. 1.2). Other features characteristic of canids are the loss of a stapedial artery and the medial position of the internal carotid artery that is situated between the entotympanic and the bone that contains the inner ear (the petrosal) for most of its course. Anteriorly the internal carotid is contained within the rostral entotympanic just before entering the braincase (Wang and Tedford, 1994). These basicranial characteristics have remained more or less stable throughout the history of canids, allowing easy identification in the fossil record when these structures are preserved.

Evolutionary History

Among the living families in the order Carnivora, the Canidae are the most ancient. The family arose in the late Eocene, when no other living families of carnivorans had yet emerged (two archaic families, Miacidae and Viverravidae, have a much older history but none survive to the present time). Furthermore, canids still maintain many features that are primitive among all carnivorans, to the extent that dog skulls are often used to illustrate a generalized mammal in zoological classrooms. Dentally, canids are closest to the ancestral morphotype of Carnivora. Canids have a relatively unreduced dental formula of 3142/3143 (numbers in sequence represent incisors, canines, premolars and molars in the upper (left half before the oblique) and the lower (right half after the oblique) teeth). These are relatively unmodified tribosphenic molars except for the morphology of the carnassials (P4, m1), which are typical of all carnivorans. In contrast, all other carnivoran families generally have a more reduced dental formula and highly modified cusp patterns.

From this mesocarnivorous (moderately carnivorous) conservative plan, canids generally evolved toward a hypercarnivorous (highly carnivorous) or hypocarnivorous (slightly carnivorous) dental pattern. In the hypercarnivorous pattern (Fig. 1.3B, D) there is a general tendency for the size of the carnassial pair to be enlarged at the expense of the molars behind (see also *Enhydrocyon*, *Aelurodon*, *Borophagus* and *Cuon* in Fig. 1.4). This modification increases the efficiency of carnassial shear. A hypocarnivorous pattern (Fig. 1.3A, C) is the opposite, with development of the grinding part of the dentition (molars) at the expense of carnassial shear (see also *Cynarctoides*, *Phlaocyon* and *Cynarctus* in Fig. 1.4). This configuration was only possible in the sister-taxa Borophaginae and Caninae, which share a bicuspid m1 talonid (Fig. 1.3C). One of the major trends in canid evolution is the repeated development of hyper- and hypocarnivorous forms (see below).

Hesperocyoninae

The subfamily Hesperocyoninae is the first major clade (a clade refers to a natural group of organisms that share a common ancestry) with a total of 28 species (Fig. 1.4). Its earliest members are species of the small fox-like form, *Hesperocyon*, that first appeared in the late Eocene (40–37 Ma) (Bryant, 1992) and became abundant in the latest Eocene. By the Oligocene (34–30 Ma), early members of four small clades of the hesperocyonines had emerged: *Paraenhydrocyon*, *Enhydrocyon*, *Osbornodon* and *Ectopocynus*. Hesperocyonines experienced their maximum diversity of 14 species during the late Oligocene (30–28 Ma), and reached their peak predatory adaptations (hypercarnivory) in the earliest Miocene with advanced species of *Enhydrocyon* and *Paraenhydrocyon*. The last species of the subfamily, *Osbornodon fricki*, became extinct in the middle Miocene (15 Ma), reaching the size of a small wolf.

With the exception of the *Osbornodon* clade, which acquired a bicuspid m1 talonid, hesperocyonines are primitively hypercarnivorous in dental adaptations with tendencies toward reduced last molars and trenchant (single cusped) talonid heels on the lower first molar. Although never reaching the extremes seen in the borophagines (see below), hesperocyonines had modest development of bone-cracking adaptations in their strong premolars. At least three lineages, all species of *Enhydrocyon* and terminal species of *Osbornodon* and *Ectopocynus*, have independently evolved their own unique array of bone-cracking teeth. Hesperocyonines did not experiment with hypocarnivory.

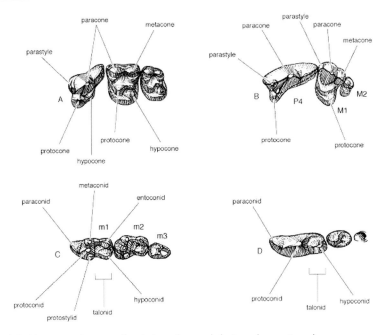

Fig. 1.3. Hypercarnivorous (b, *Aelurodon* and d, *Euoplocyon*) and hypocarnivorous (a, *Phlaocyon* and c, *Cynarctus*) dentitions. In hypercarnivorous forms, the upper cheek teeth (B) tend to emphasize the shearing part of the dentition with an elongated and narrow P4, an enlarged parastyle on a transversely elongated M1, and a reduced M2. On the lower teeth (D), hypercarnivory is exemplified by a trenchant talonid due to the increased size and height of the hypoconid at the expense of the entoconid (reduced to a narrow and low ridge), accompanied by the enlargement of the protoconid at the expense of the metaconid (completely lost in *Euoplocyon*) and the elongation of the trigonid at the expense of the talonid. In hypocarnivorous forms, on the other hand, the upper teeth (A) emphasize the grinding part of the dentition with a shortened and broadened P4 (sometimes with a hypocone along the lingual border), a reduced parastyle on a quadrate M1 that has additional cusps (e.g. a conical hypocone along the internal cingulum) and cuspules, and an enlarged M2. The lower teeth (C) in hypocarnivorous forms possess a basined (bicuspid) talonid on m1 enclosed on either side by the hypoconid and entoconid that are approximately equal in size. Other signs of hypocarnivory on the lower teeth include widened lower molars, enlarged metaconids, and additional cuspules such as a protostylid.

Borophaginae

From the primitive condition of a trenchant talonid heel on the lower first molar seen in the hesperocyonines, borophagines and canines shared a basined (bicuspid) talonid acquired at the very beginning of their common ancestry (Fig. 1.3C). Along with a more quadrate upper first molar with its hypocone, the basined talonid establishes an ancestral state from which all subsequent forms were

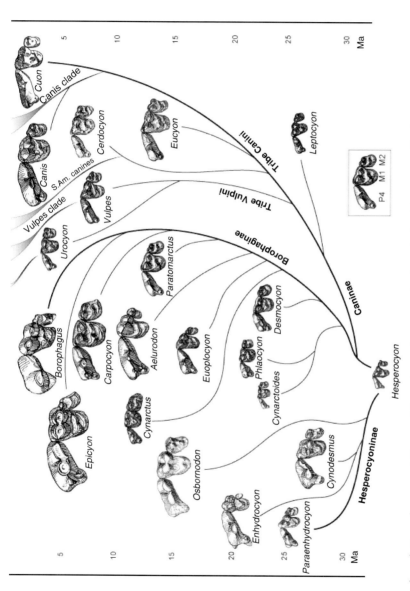

Fig. 1.4. Dental evolution of representative canids as shown in upper cheek teeth (P4–M2). Generally the most advanced species in each genus is chosen to enhance a sense of dental diversity. Species in the Hesperocyoninae are: *Hesperocyon gregarius*; *Paraenhydrocyon josephi*; *Cynodesmus martini*; *Enhydrocyon crassidens*; and *Osbornodon fricki*. Species in the Borophaginae are: *Cynarctoides acridens*; *Phlaocyon marslandensis*; *Desmocyon thomsoni*; *Cynarctus crucidens*; *Euoplocyon brachygnathus*; *Aelurodon stirtoni*; *Paratomarctus temerarius*; *Carpocyon webbi*; *Epicyon haydeni*; and *Borophagus diversidens*. Species in the Caninae are: *Leptocyon gregorii*; *Vulpes stenognathus*; *Urocyon minicephalus*; *Cerdocyon thous*; *Eucyon davisi*; *Canis dirus*; and *Cuon alpinus*. All teeth are scaled to be proportional to their sizes.

derived. Such a dental pattern proved to be very versatile and can readily be adapted toward either a hyper- or hypocarnivorous type of dentition, both of which were repeatedly employed by both borophagines and canines (Fig. 1.4).

The history of the borophagines also begins with a small fox-like form, *Archaeocyon*, in the late Oligocene. Contemporaneous with larger and more predatory hesperocyonines, these early borophagines in the late Oligocene and early Miocene tended to be more omnivorous (hypocarnivorous) in their dental adaptations, such as *Oxetocyon*, *Otarocyon* and *Phlaocyon*. One extreme case, *Cynarctoides* evolved selenodont-like molars as in modern artiodactyles, a rare occurrence of herbivory among carnivorans. These early borophagines are generally no larger than a raccoon, which is probably a good ecological model for some borophagines at a time when procyonids had yet to diversify.

After some transitional forms in the early Miocene, such as *Cormocyon* and *Desmocyon*, borophagines achieved their maximum ecological and numerical (i.e. species) diversity in the middle Miocene, with highly omnivorous forms, such as *Cynarctus*, that were almost ursid-like, as well as highly predatory forms, such as *Aelurodon*, that were a larger version of the living African hunting dog *Lycaon*. By then, borophagines had acquired their unique characteristics of a broad muzzle, a bony contact between premaxillary and frontal, multicuspid incisors, and an enlarged parastyle on the upper carnassials (modified from an enlargement of the anterior cingulum).

By the end of the Miocene, borophagines had evolved another lineage of omnivores, although only modestly in that direction, in the form of *Carpocyon*. Species of *Carpocyon* are mostly the size of jackals to small wolves. At the same time, the emergence of the genus *Epicyon* from a *Carpocyon*-like ancestor marked another major clade of hypercarnivorous borophagines. The terminal species of *Epicyon*, *E. haydeni*, reached the size of a large bear and holds the record as the largest canid ever to have lived. Closely related to *Epicyon* is *Borophagus*, the terminal genus of the Borophaginae. Both *Epicyon* and *Borophagus* are best known for their massive P4 and p4 in contrast to the diminutive premolars in front. This pair of enlarged premolars is designed for cracking bones, mirroring similar adaptations by hyaenids in the Old World. Advanced species of *Borophagus* survived most of the Pliocene but became extinct near the beginning of the Pleistocene.

Caninae

As in the hesperocyonines and borophagines, a small fox-sized species of *Leptocyon* is the earliest recognized member of the subfamily Caninae. Besides sharing a bicuspid talonid of m1 and a quadrate M1 with the borophagines, *Leptocyon* is also characterized by a slender rostrum and elongated lower jaw, and correspondingly narrow and slim premolars, features that are inherited in all subsequent canines. It first appeared in the early Oligocene and persisted into the late Miocene. Throughout its long existence (no other canid genus had as long a duration), facing intense competition from the larger and diverse hesperocyonines and borophagines, *Leptocyon* generally remains small and inconspicuous, never having more than two or three species at a time.

By the latest Miocene, fox-sized niches are widely available in North America, left open by extinctions of all small borophagines. The true fox clade, tribe Vulpini, emerges at this time and undergoes a modest diversification to initiate primitive species of both *Vulpes* and *Urocyon* (and their extinct relatives). The North American Pliocene record of *Vulpes* is quite poor. Fragmentary materials from early Blancan indicate the presence of a swift fox-like form in the Great Plains. *Vulpes* species were widespread and diverse in Eurasia during the Pliocene (see Qiu and Tedford, 1990), resulting from an immigration event independent from that of the *Canis* clade. Red fox (*Vulpes vulpes*) and Arctic fox (*Vulpes lagopus*) appeared in North America only in the late Pleistocene, evidently as a result of immigration back to the New World.

Preferring more wooded areas, the grey fox *Urocyon* has remained in southern North America and Middle America. Records of the grey fox clade indicate a more or less continuous presence in North America throughout its existence, with intermediate forms leading to the living species *U. cinereoargenteus*. Morphologically, the living African bat-eared fox *Otocyon* is closest to the *Urocyon* clade, although molecular evidence suggests that the bat-eared fox may lie at the base of the fox clade or even lower (Geffen *et al.*, 1992; Wayne *et al.*, 1997). If the morphological evidence has been correctly interpreted, then the bat-eared fox must represent a Pliocene immigration event to the Old World independent of other foxes. A transitional form, *Protocyon*, occurs in southern Asia and Africa in the early Pleistocene.

Advanced members of the Caninae, tribe Canini, first occur in the middle Miocene (9–12 Ma) in the form of a transitional taxon *Eucyon*. As a jackal-sized canid, *Eucyon* is mostly distinguished from the Vulpini in an expanded paroccipital process and enlarged mastoid process, and in the consistent presence of a frontal sinus. The latter character initiates a series of transformations in the Tribe Canini culminating in the elaborate development of the sinuses and a domed skull in *Canis lupus*. By latest Miocene time, species of *Eucyon* have appeared in Europe (Rook, 1992) and by the early Pliocene in Asia (Tedford and Qiu, 1996). The North American records all pre-date the European ones, suggesting a westward dispersal of this form.

Arising from about the same phylogenetic level as *Eucyon* is the South American clade (subtribe Cerdocyonina). Morphological and molecular evidence generally agrees that living South American canids, the most diverse group of canids on a single continent, belong to a natural group of their own. The South American canids are united by morphological characters such as a long palate, a large angular process of the jaw with a widened scar for attachment of the inferior branch of the medial pterygoid muscle, and a relatively long base of the coronoid process (Tedford *et al.*, 1995). By the close of the Miocene, certain fragmentary materials from southern United States and Mexico indicate that taxa assignable to *Cerdocyon* (Torres and Ferrusquía-Villafranca, 1981) and *Chrysocyon* occur in North America. The presence of these advanced taxa in the North

American late Miocene predicts that ancestral stocks of many of the South American canids may have been present in southern North America or Middle America. They appear in the South American fossil record shortly after the formation of the Isthmus of Panama in the Pliocene, around 3 Ma (Berta, 1987). The earliest records are *Pseudalopex* and its close relative *Protocyon*, an extinct large hypercarnivore, from the Plio-Pleistocene (around 2.5–1.5 Ma) of Argentina. By the latest Pleistocene (50,000–10,000 years ago), most living species or their close relatives had emerged, along with the extinct North American dire wolf, *Canis dirus*. By the end of the Pleistocene, all large, hypercarnivorous canids of South America (*Protocyon, Theriodictis*) as well as *Canis dirus* had become extinct.

The *Canis* clade within the tribe Canini, the most advanced group in terms of large size and hypercarnivory, arises near the Miocene–Pliocene boundary between 5 and 6 Ma in North America. A series of jackal-sized ancestral species of *Canis* thrived in the early Pliocene, such as *C. ferox, C. lepophagus* and other undescribed species. At about the same time, the first records of canids begin to appear in the European late Neogene: *Canis cipio* in the late Miocene of Spain (Crusafont-Pairó, 1950), *Eucyon monticinensis* in the latest Miocene of Italy (Rook, 1992), the earliest raccoon-dog *Nyctereutes donnezani* and the jackal-sized *Canis adoxus* in the early Pliocene of France (Martin, 1973; Ginsburg, 1999). The enigmatic *C. cipio*, only represented by parts of the upper and lower dentition, may pertain to a form at the *Eucyon* level of differentiation rather than truly a species of *Canis*.

The next phase of *Canis* evolution is difficult to track. The newly arrived *Canis* in Eurasia underwent an extensive radiation and range expansion in the late Pliocene and Pleistocene, resulting in multiple, closely related species in Europe, Africa and Asia. To compound this problem, the highly cursorial wolf-like *Canis* species apparently belong to a circum-arctic fauna that undergoes expansions and contractions with the fluctuating climate. Hypercarnivorous adaptations are common in the crown-group of species, especially in the Eurasian middle latitudes and Africa. For the first time in canid history, phylogenetic studies cannot be satisfactorily performed on forms from any single continent because of their Holarctic distribution and faunal intermingling between the New and Old Worlds. Nevertheless some clades were localized in different parts of Holarctica. The vulpines' major centre of radiation was in the Old World. For the canines, North America remained a centre through the Pliocene producing the coyote as an endemic form. A larger radiation yielding the wolves, dhole, African hunting dog and fossil relatives took place on the Eurasian and African continents. During the Pleistocene elements of the larger canid fauna invaded mid-latitude North America – the last invasion of which was the appearance of the grey wolf south of the glacial ice sheets in the latest Pleistocene (about 100,000 years ago).

Phylogenetic Relationships

As mentioned above, there is strong fossil evidence about the antiquity of the family Canidae. This basal placement within the suborder Caniformia is increasingly born out by molecular data in recent years, such as DNA–DNA hybridization of single copy DNA, mitochondrial DNA sequence studies, and recently studies of DNA sequences from nuclear genes (Vrana *et al.*, 1994; Slattery and Brien, 1995; Flynn and Nedbal, 1998; Murphy *et al.*, 2001; Wang *et al.*, 2004a), although molecular clock calculations tend to place the divergence time somewhat older, around 50 Ma (Wayne *et al.*, 1989), than is estimated from fossil evidence.

Phylogenetic (genealogical) relationships are traditionally inferred by analysis of the morphological characters, but molecular data are increasingly playing important, sometimes controversial, roles in the detection of evolutionary relationships. However, in the case of canids that have a substantial history known by fossil records only, morphology is still the only way to allow a comprehensive view of their entire history.

For the two extinct subfamilies, Hesperocyoninae and Borophaginae, we have performed an exhaustive analysis of the entire fossil records (Wang, 1994; Wang *et al.*, 1999) and their relationships are roughly shown in Fig. 1.4. For the subfamily Caninae, we have nearly finished a similar study of monographic revisions that deals with the entire fossil history of the canines in North America (Tedford *et al.*, in prep.). As a part of this larger effort to lay down a phylogenetic framework, Tedford *et al.* (1995) performed a cladistic analysis of living canids on morphological grounds. The result is a nearly fully resolved relationship based on an 18 taxa by 57 characters matrix at the generic level. This relationship recognizes three monophyletic clades in the canines: the fox group (tribe Vulpini), the South American canine group, and the wolf group containing hypercarnivorous forms (the latter two form the tribe Canini) (Fig. 1.5, left).

Molecular studies of canid relationships range from investigations in comparative karyology, allozyme electrophoresis, mitochondrial DNA, to microsatellite loci (Wayne and Brien, 1987; Wayne *et al.*, 1987a, b, 1997; Geffen *et al.*, 1992; Bruford and Wayne, 1993; Girman *et al.*, 1993; Gottelli *et al.*, 1994; Vilà *et al.*, 1997, 1999). Trees derived from the mtDNA are based on the widest possible sample of taxa and are better studied than nuclear DNA (Fig. 1.5, right) (e.g. Wayne *et al.*, 1997). Overall, molecular studies tend to place the foxes near the basal part, the South American canines in the middle, and the wolves and hunting dogs toward the terminal branches, a pattern that is consistent with the morphological tree. The detailed arrangements, however, differ in a number of ways. The foxes are generally in a paraphyletic arrangement (falling at the stem parts of the tree) in contrast to a monophyletic clade (a natural group that contains ancestors and *all* descendants) in the morphological tree. The grey fox and bat-eared fox are placed at the base despite their highly advanced dental morphology compared to other foxes. Similarly, South American canines are no longer monophyletic under molecular analysis but form at least two paraphyletic branches. A glaring discrepancy is the Asiatic raccoon dog being allied to the foxes in the molecular analysis despite its numerous morphological characters shared with some South American forms. Finally, molecular data suggest independent origins for the Asiatic and African hunting dogs in contrast to a sister

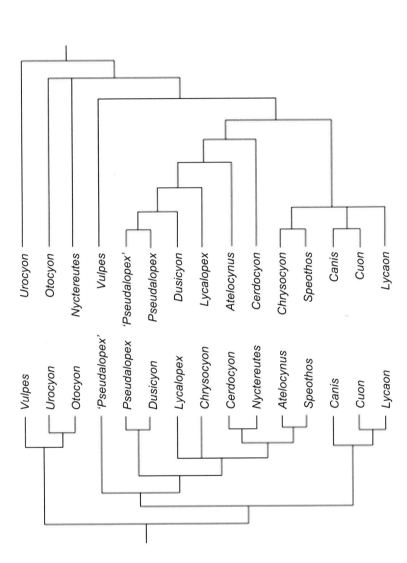

Fig. 1.5. Contrasting canine relationships from recent morphological (left) (Tedford et al., 1995) and molecular studies (right) (Wayne et al., 1997).

relationship in the morphological tree supported by a large number of characters related to hypercarnivory.

Not surprisingly, there are increased agreements between the molecular and morphological results when the two data sets are combined in a total evidence analysis (Wayne *et al.*, 1997; Zrzavý and Ricánková, 2004). Under such conditions, the South American canines (except *Nyctereutes*) become monophyletic, as does the clade including the wolf, dhole and African hunting dog. Although mitochondrial genes are known to evolve relatively quickly and thus are suitable for analysis on groups that have more recent ancestry, genetic variations in the mitochondria are only a tiny fraction of the total genome and the gene trees may not reveal the true phylogeny. Nuclear genes, with their vast information content, have the potential of revealing the true relationships, especially for groups of longer history. Works are underway to search for the most suitable part of the genome with just the right rate of evolution, neither too slow to offer much insight about change nor too fast to obscure true relationships. Recently Selenocysteine tRNA (Cf TRSP) and RNase P RNA (Cf RPPH1) genes have been shown to be promising (Bardeleben *et al.*, 2005), these tend to place the South American forms in a clade and are consistent with morphological results.

Evolutionary Trends

As a very successful group of predators, canids are known for their outstanding cursoriality, the ability to run fast and over long distances, and for their social (pack) hunting that requires complex collaborative behaviours. This combination of long-distance, relay-style running and social hunting to bring down prey together is a successful strategy in catching larger prey. In this regard, only the hyaenids are comparably equipped, whereas the felids may have reached similar running capabilities but more often hunt alone. The increased cursoriality in canids is generally correlated with a similarly increased running ability by their prey (various groups of ungulates), which are in turn related to the progressively more open grassland environments in the late Cenozoic.

Running and posture

Increased ability of running is often manifested in elongated and slender limbs, reduction of digits, and more erect standing posture. In addition to the overall elongation of the limbs, the distal segments (radius–ulna, tibia–fibular and metapodials) in particular tend to elongate more than the proximal segments (humerus and femur). The reduction of digits in carnivorans is usually confined to the decline of the digit I to result in a functionally four-digit hand and foot, in contrast to the far more dramatic digit reduction in ungulates such as the horses, because of the need for grasping by the hands in carnivorans. In all cursorial carnivorans, the standing posture is usually digitigrade with the metapodials lifted from the ground in contrast to the primitively plantigrade posture with proximal ends of the metapodials still touching the ground. All of the above limb modifications are a common strategy to increase the length of stride and decrease the weight in distal portions of the limbs.

Although fossil postcranial skeletons are often less abundant than cranial and dental materials, particularly those that are associated with dental materials to be accurately identified in taxonomy, we are in possession of enough skeletons in a few taxa in most major clades to permit a reasonable grasp of the general evolutionary trends. Starting from the beginning of the Canidae, *Hesperocyon* has already exhibited an initial stage of cursoriality with moderate lengthening of limbs and a semi-digitigrade posture (Wang, 1993). Within the subfamily Caninae, this trend steadily progressed from the basal genus *Leptocyon*, which has increasingly slender limbs, to *Vulpes*, which has lost the entepicondylar foramen on the distal humerus (a sign of further slendering) and substantially reduced digit I, and to the *Canis* clade that carries these trends to a greater extreme as shown by living canids (Tedford *et al.*, 1995).

Social hunting

Hunting behaviour is generally not preserved in fossil records and canids' pack hunting behaviour can only be approached in an indirect way. By correlating certain skull morphology and body size, Van Valkenburgh *et al.* (2003) suggested that some hypercarnivorous borophagines may have acquired social hunting skills, although likely independently from those in the canines. Social hunting in the Caninae is mostly confined to the *Canis* clade (Macdonald *et al.*, 2004) and thus has presumably arisen in the last few million years of canid history (see earlier section on Caninae history).

Diet and body size

Finally, throughout their history canids displayed remarkable flexibility and diversity in adaptations to different diets and preys. Small, generalized canids have the ability to evolve toward either hypocarnivorous or hypercarnivorous directions, depending on opportunities (see Fig. 1.3 for dental adaptations). Hypocarnivorous forms tend to remain small- to medium-sized, and sometimes reduce their body size. The hypercarnivorous direction, however, often leads to larger body size, possibly as a result of energetic requirements (Carbone *et al.*, 1999). As in the history of hyaenids, large, hypercarnivorous canids frequently developed strong premolars that are capable of cracking bones as an additional source of protein (Werdelin, 1989). Van Valkenburgh *et al.* (2004) suggested that such a correlated increase in body size and hypercarnivory acts like an evolutionary ratchet – once a certain lineage is launched toward increased predation, a larger body size often evolves over time.

Acknowledgements

We thank Per Jensen for his invitation to write this contribution. This research is funded in part by grants from the National Science Foundation (DEB 9420004; 9707555).

References

Bardeleben, C., Moore, R.L. and Wayne, R.K. (2005) Isolation and molecular evolution of the Selenocysteine tRNA (Cf TRSP) and RNase P RNA (Cf RPPH1) genes in the dog family, Canidae. *Molecular Biology and Evolution* 22, 347–359.

Berta, A. (1987) Origin, diversification, and zoogeography of the South American Canidae. In: Patterson, B.D. and Timm, R.M. (eds) *Studies in Neotropical Mammalogy: Essays in Honor of Philip Hershkovitz.* Field Museum of Natural History, Chicago, Illinois, pp. 455–471.

Berta, A. (1988) Quaternary evolution and biogeography of the large South American Canidae (Mammalia: Carnivora). *University of California Publications of Geological Science* 132, 1–149.

Bruford, M.W. and Wayne, R.K. (1993) Microsatellites and their application to population genetic studies. *Current Biology* 3, 939–943.

Bryant, H.N. (1992) The Carnivora of the Lac Pelletier Lower Fauna (Eocene: Duchesnean), Cypress Hills Formation, Saskatchewan. *Journal of Paleontology* 66, 847–855.

Carbone, C., Mace, G.M., Roberts, S.C. and Macdonald, D.W. (1999) Energetic constraints on the diet of terrestrial carnivores. *Nature* 402, 286–288.

Crusafont-Pairó, M. (1950) El primer representante del género *Canis* en el Pontiense eurasiatico (*Canis cipio* nova sp.). *Boletin de la Real Sociedad Española de Historia Natural (Geologia)* 48, 43–51.

Flynn, J.J. and Nedbal, M.A. (1998) Phylogeny of the Carnivora (Mammalia): congruence vs. incompatibility among multiple data sets. *Molecular Phylogenetics and Evolution* 9, 414–426.

Geffen, E., Mercure, A., Girman, D.J., Macdonald, D.W. and Wayne, R.K. (1992) Phylogenetic relationships of the fox-like canids: mitochondrial DNA restriction fragment, site and cytochrome *b* sequence analyses. *Journal of Zoology* 228, 27–39.

Ginsburg, L. (1999) Order Carnivora. In: Rössner, G.E. and Heissig, K. (eds) *The Miocene Land Mammals of Europe.* Verlag Dr. Friedrich Pfeil, München, Germany, pp. 109–148.

Girman, D.J., Kat, P.W., Mills, G., Ginsberg, J., Fanshaw, J., Fitzgibbon, C., Borner, M., Wilson, V., Laurenson, K. and Wayne, R.K. (1993) A genetic and morphological analysis of the African wild dog (*Lycaon pictus*). *Journal of Heredity* 84, 450–459.

Gottelli, D., Sillero-Zubiri, C., Applebaum, G.D., Roy, M.S., Girman, D.J., Garcia-Moreno, J., Ostrander, E.A. and Wayne, R.K. (1994) Molecular genetics of the most endangered canid: the Ethiopian wolf, *Canis simensis. Molecular Ecology* 3, 277–290.

Hunt, R.M., Jr (1974) The auditory bulla in Carnivora: an anatomical basis for reappraisal of carnivore evolution. *Journal of Morphology* 143, 21–76.

Leonard, J.A., Wayne, R.K., Wheeler, J., Valadez, R., Guillén, S. and Vilà, C. (2002) Ancient DNA evidence for old world origin of new world dogs. *Science* 298, 1613–1616.

Linnaeus, C. (1758) *Systema naturae per regna tria naturae, secundum classes, ordines, genera, species cum characteribus, differentiis, synonymis, locis. Editio Decima, 1758 (12th edition of Linaeus 1758).* Societatis Zoologicae Germanicae.

Macdonald, D.W., Creel, S. and Mills, M.G.L. (2004) Canid society. In: Macdonald, D.W. and Sillero-Zubiri, C. (eds) *The Biology and Conservation of Wild Canids.* Oxford University Press, Oxford, UK, pp. 85–106.

Martin, R. (1973) Trois nouvelles espèces de Caninae (Canidae, Carnivora) des gisements plio-villafranchiens d'Europe. *Documents des Laboratoires de Géologie de la Faculté des Sciences de Lyon* 57, 87–96.

Murphy, W.J., Eizirik, E., O'Brien, S.J., Madsen, O., Scally, M., Douady, C.J., Teeling, E., Ryder, O.A., Stanhope, M.J., de Jong, W.W. and Springer, M.S. (2001) Resolution of the early placental mammal radiation using Bayesian phylogenetics. *Science* 294, 2348–2351.

Pennisi, E. (2002) A shaggy dog history. *Science* 298, 1540–1542.

Qiu, Z.-X. and Tedford, R.H. (1990) A Pliocene species of *Vulpes* from Yushe, Shanxi. *Vertebrata PalAsiatica* 28, 245–258.

Rook, L. (1992) '*Canis*' *monticinensis* sp. nov., a new Canidae (Carnivora, Mammalia) from the late Messinian of Italy. *Bolletino della Società Paleontologica Italiana* 31, 151–156.

Savolainen, P., Zhang, Y.-P., Luo, J., Lundeberg, J. and Leitner, T. (2002) Genetic evidence for an East Asian origin of domestic dogs. *Science* 298, 1610–1613.

Slattery, J.P. and Brien, S.J.O. (1995) Molecular phylogeny of the red panda (*Ailurus fulgens*). *Journal of Heredity* 86, 413–422.

Tedford, R.H. (1978) History of dogs and cats: a view from the fossil record. *Nutrition and Management of Dogs and Cats.* Ralston Purina Co., St Louis, Missouri, chap. M23.

Tedford, R.H. and Qiu, Z.-X. (1996) A new canid genus from the Pliocene of Yushe, Shanxi Province. *Vertebrata PalAsiatica* 34, 27–40.

Tedford, R.H., Taylor, B.E. and Wang, X. (1995) Phylogeny of the Caninae (Carnivora: Canidae): the living taxa. *American Museum Novitates* 3146, 1–37.

Tedford, R.H., Wang, X. and Taylor, B.E. (in prep.) Phylogenetic systematics of the North American fossil Caninae (Carnivora: Canidae). *Bulletin of the American Museum of Natural History.*

Torres, R.V. and Ferrusquía-Villafranca, I. (1981) *Cerdocyon* sp. nov. A (Mammalia, Carnivora) en Mexico y su significacion evolutiva y zoogeografica en relacion a los canidos sudamericanos. In: *Anais II Congresso Latino-Americano de Paleontologia*, Porto Alegre, Brazil, pp. 709–719.

Van Valkenburgh, B., Sacco, T. and Wang, X. (2003) Pack hunting in Miocene borophagine dogs: evidence from craniodental morphology and body size. In: Flynn, L.J. (ed.) *Vertebrate Fossils and Their Context: Contributions in Honor of Richard H. Tedford. Bulletin of the American Museum of Natural History.* American Museum of Natural History, New York, pp. 147–162.

Van Valkenburgh, B., Wang, X. and Damuth, J. (2004) Cope's Rule, hypercarnivory, and extinction in North American Canids. *Science* 306, 101–104.

Vilà, C., Savolainen, P., Maldonado, J.E., Amorim, I.R., Rice, J.E., Honeycutt, R.L., Crandall, K.A., Lundeberg, J. and Wayne, R.K. (1997) The domestic dog has an ancient and genetically diverse origin. *Science* 276, 1687–1689.

Vilà, C., Amorim, I.R., Leonard, J.A., Posada, D., Castroviejo, J., Petrucci-Fonseca, F., Crandall, K.A., Ellegren, H. and Wayne, R.K. (1999) Mitochondrial DNA phylogeography and population history of the grey wolf *Canis lupus*. *Molecular Ecology* 8, 2089–2103.

Vrana, P.B., Milinkovitich, M.C., Powell, J.R. and Wheeler, W.C. (1994) Higher level relationships of the arctoid Carnivora based on sequence data and 'total evidence'. *Molecular Phylogenetics and Evolution* 3, 47–58.

Wang, X. (1993) Transformation from plantigrady to digitigrady: functional morphology of locomotion in *Hesperocyon* (Canidae: Carnivora). *American Museum Novitates* 3069, 1–23.

Wang, X. (1994) Phylogenetic systematics of the Hesperocyoninae (Carnivora: Canidae). *Bulletin of the American Museum of Natural History* 221, 1–207.

Wang, X. and Tedford, R.H. (1994) Basicranial anatomy and phylogeny of primitive canids and closely related miacids (Carnivora: Mammalia). *American Museum Novitates* 3092, 1–34.

Wang, X. and Tedford, R.H. (1996) Canidae. In: Prothero, D.R. and Emry, R.J. (eds) *The Terrestrial Eocene–Oligocene Transition in North America, Pt. II: Common Vertebrates of the White River Chronofauna.* Cambridge University Press, Cambridge, pp. 433–452.

Wang, X., Tedford, R.H. and Taylor, B.E. (1999) Phylogenetic systematics of the Borophaginae (Carnivora: Canidae). *Bulletin of the American Museum of Natural History* 243, 1–391.

Wang, X., Tedford, R.H., Van Valkenburgh, B. and Wayne, R.K. (2004a) Ancestry: Evolutionary history, molecular systematics, and evolutionary ecology of Canidae. In: Macdonald, D.W. and Sillero-Zubiri, C. (eds) *The Biology and Conservation of Wild Canids.* Oxford University Press, Oxford, UK, pp. 39–54.

Wang, X., Tedford, R.H., Van Valkenburgh, B. and Wayne, R.K. (2004b) Phylogeny, classification, and evolutionary ecology of the Canidae. In: Sillero-Zubiri, C., Hoffmann, M. and Macdonald, D.W. (eds) *Canids: Foxes, Wolves, Jackals and Dogs. Status Survey and Conservation Action Plan.* IUCN/SSC Canid Specialist Group, The World Conservation Union, Gland, Switzerland, pp. 8–20.

Wayne, R.K. and Brien, S.J.O. (1987) Allozyme divergence within the Canidae. *Systematic Zoology* 36, 339–355.

Wayne, R.K., Nash, W.G. and Brien, S.J.O. (1987a) Chromosomal evolution of the Canidae. I. Species with high diploid numbers. *Cytogenetics and Cell Genetics* 44, 123–133.

Wayne, R.K., Nash, W.G. and Brien, S.J.O. (1987b) Chromosomal evolution of the Canidae. II. Divergence from the primitive carnivore karyotype. *Cytogenetics and Cell Genetics* 44, 134–141.

Wayne, R.K., Benveniste, R.E., Janczewski, D.N. and O'Brien, S.J. (1989) Molecular and biochemical evolution of the Carnivora. In: Gittleman, J.L. (ed.) *Carnivore Behavior, Ecology, and Evolution.* Cornell University Press, Ithaca, New York, pp. 465–494.

Wayne, R.K., Geffen, E., Girman, D.J., Koepfli, K.-P., Lau, L.M. and Marshall, C.R. (1997) Molecular systematics of the Canidae. *Systematic Zoology* 46, 622–653.

Werdelin, L. (1989) Constraint and adaptation in the bone-cracking canid *Osteoborus* (Mammalia: Canidae). *Paleobiology* 15, 387–401.

Zrzavý, J. and Ricánková, V. (2004) Phylogeny of recent Canidae (Mammalia, Carnivora): relative reliability and utility of morphological and molecular datasets. *Zoologica Scripta* 33, 311–333.

Domestication of Dogs

Peter Savolainen

Introduction

The dog is perhaps the most fascinating of the domestic animals. According to the available knowledge, it was the first domestic animal, and its wild ancestor, the wolf, was probably domesticated by mobile hunter-gatherers rather than by settled farmers, contrary to probably all other domestic animals (Clutton-Brock, 1995). The dog is morphologically very diverse, both in size and shape. In the range of size it is the most diverse mammal species, with the close to 100-fold difference in weight between the chihuahua and the great dane being the most prominent example. Dogs and humans share a number of social and behavioural signals facilitating communication. Many of these are shared with the wolf and have undoubtedly had significance in the domestication of the wolf, facilitating the first contacts, but have also been further selected during the process of domestication (Houpt and Willis, 2001). This gives the dog a behaviour which appeals to many of us more than that of other domestic animals, explaining its status as man's best friend.

Despite this special status, until recently surprisingly little has been known about the history of the dog. While the origin from the wolf has been the generally accepted theory, there have been virtually no other details known about the first origin of the dog, or about how the immense morphological variation, and the hundreds of dog breeds, have developed. Since there is great interest in this subject, numerous texts have been written, and there have been many theories about how, where and when the dog originated, but there have been few facts to build on, making reading about this subject an often frustrating experience.

© CAB International 2007. *The Behavioural Biology of Dogs*
(ed. P. Jensen)

However, in the last 10 years, genetic analyses, based on sequence analysis of mitochondrial DNA (mtDNA), have provided a large step toward unravelling the origin and early history of the domestic dog. Thus, the origin of the dog from the wolf has been shown with great certainty, and for the first time there is relatively firm proof to suggest a geographic origin of the dog. While more studies are needed to definitely establish the facts, the sum of the available evidence indicates that the domestic dog has a single origin from East Asian wolves, some 15,000 years ago.

Origin from the Wolf

There is now overwhelming evidence that the wild ancestor of the domestic dog is the wolf. An origin primarily from the wolf has been the main theory from the start of evolutionary theory building, but it has been hypothesized that also other canids, such as the jackal or coyote, could have contributed to the forming of the dog, explaining some of the morphological and behavioural variation among domestic dogs. However, with time, numerous studies of physical and behavioural characteristics, and later also molecular genetic markers, both nuclear markers and mtDNA, have built an increasingly stronger case for the wolf origin (Wayne, 1993; Clutton-Brock, 1995). While not disproving some degree of contribution from other species, the total collection of evidence is strong in favour of the wolf as the only ancestor of the dog. The evidence from mtDNA gives a consistent picture: among more than 1000 analysed dogs, all have mtDNA sequences which are much closer to those of wolves than of the other wild canids. However, mtDNA is maternally inherited and can therefore only describe the history of females, and while studies of nuclear markers consistently support an origin from the wolf, studies have so far been relatively limited, leaving a possibility for a small contribution on the male side from, for example, the jackal. Large-scale population studies of nuclear markers, for example the Y chromosome, will surely give the final answer within a few years.

The Archaeological Evidence

Archaeological finds related to dogs have been found at Mesolithic sites in Europe, Asia and America, the earliest going back to at least 14,000 BP (years before present), indicating that the dog was the first domestic animal, several thousand years ahead of the first farm animals (Clutton-Brock, 1995). The earliest finds believed to be from domestic dogs are from Central and Eastern Europe and the Middle East: a single jaw from 14,000 BP in Germany (Nobis, 1979), two skulls from 13,000–17,000 BP in western Russia (Sablin and Khlopachev, 2002) and an assemblage of small canids from 12,000 BP in the Middle East (Tchernov and Valla, 1997). The earliest finds from East Asia and North America are younger, from 7500 (Li, 1990) and 8500 BP (Clutton-Brock and Noe-Nygaard,

1990), respectively. However, the archaeological evidence is problematic to interpret. It is difficult to discriminate between species among canid remains, for example between small wolves and domestic dogs (Olsen, 1985), the dating of the finds may be obscured by, for example, disturbed layers, and archaeological excavations have not been evenly performed in different parts of the world. Thus, the fact that the oldest remains of dogs have been found in the relatively thoroughly excavated regions of Europe and the Middle East could be attributed to a greater chance of finding remains there than in less explored regions. The morphology and size of early archaeological dog remains suggest an origin from one of the small South Asian wolf subspecies rather than from the large North Eurasian and North American wolves, and the spread of the wolf, which has been limited to Eurasia and North America, excludes the possibility of an African origin (Clutton-Brock, 1995). One osteological feature of the jaw, diagnostic for dogs, is found also among Chinese wolves, but rarely in other wolves, indicating an origin from East Asia (Olsen, 1985).

Thus, the archaeological record has not offered any distinct clues even to the most basic historic questions, such as where, when and how many times the wolf was domesticated into the domestic dog. Based on the archaeological evidence, there have been two main theories for the origin of the domestic dog. The Middle East has been proposed, based on the earliest finds of dogs being found in Europe and the Middle East, and the small size of the local species of wolf. An origin from several different wolf populations has been suggested, based on the widespread occurrence of early finds of dog remains (Clutton-Brock, 1995). This could also to some extent offer an explanation for the extreme morphological variation among dog breeds. There is also very little known about the early history of different dog breeds, and even for the large number of breeds which have been developed in the last few hundred years, the exact origin is mostly unknown (Clutton-Brock, 1995).

Population Genetic Studies

Thus, for a long time research did not bring much new knowledge about the history of the dog, except the early molecular biological studies which could bring growing certainty that the dog originates from the wolf rather than from other canids, and the sporadic addition of new archaeological finds of early dogs, changing the geographic location for their oldest known remains. Since the classical instruments for this kind of question, archaeology and palaeontology, failed to provide any detailed information, population genetic studies was applied relatively soon after the techniques for large-scale DNA-sequencing studies became available. The studies have, so far, mainly been performed on mtDNA, because the genetic variation is higher than for nuclear DNA, especially in the noncoding part, the control region (CR), giving much genetic information from analysis of a relatively short DNA sequence. mtDNA is maternally inherited and a drawback is therefore that the mtDNA-studies can only describe the history of the female dogs, the history of the males being only indirectly monitored. An advantage on the other side, shared with the Y chromosome, compared to nuclear autosomal DNA, is that the mitochondrial genome does not recombine, which implies that the whole molecule is inherited as an entity through the generations, making the interpretation of the data straightforward.

The first three major population genetic studies of mtDNA from dogs and wolves were published in 1996 and 1997: 261 base pairs of the mtDNA CR were analysed in 140 domestic dogs representing 67 breeds, mostly European but also a few from Asia, Africa and America, and 162 wolves from throughout Europe, Asia and North America (Vila et al., 1997), and 970 and 670 base pairs, respectively, of the CR were analysed in a total of 128 dogs, representing mainly Japanese breeds, and 19 wolves (Okumura et al., 1996; Tsuda et al., 1997). These studies showed that, in phylogenetic analyses, the domestic dog mtDNA sequences were distributed into six distinct groups (clades) interspersed by wolf sequences, showing that the dog originates from several female wolf lines, at least as many as the number of clades (Fig. 2.1). Importantly, all of the 268 studied domestic dogs had mtDNA sequences which were closer to those of wolves than of other canids. Thus, there was no indication of a genetic contribution from any other canid than the wolf to the domestic dog population. However, the geographic location of the domestication of the wolf, or any detailed conclusions about the history of specific dog breeds, could not be determined based on these relatively limited data sets. The time for the origin of the dog was estimated, based on the largest genetic distance of the sequences in the largest phylogenetic clade of dog sequences, giving a possible date of 135,000 years. However, as shown below, the mtDNA data can be interpreted differently, resulting in a date more in agreement with that obtained from the archaeological record.

A Single Geographic Origin

The greatest step so far in the study of the origin of dogs was taken in 2002 with the publication of two articles, one study of the present-day dog population throughout the world (Savolainen et al., 2002) and one of pre-European archaeological remains of American dogs (Leonard et al., 2002). Leonard et al. showed that domestic dogs on all continents have a common origin from a single gene pool, and the study by Savolainen et al. suggested that this gene pool originated just once, somewhere in East Asia.

There are principally two approaches possible for elucidating the place of domestication of the wolf, using mtDNA analysis. The most straightforward approach would be to study the mtDNA sequences of today's domestic dog population and compare these with sequences from a worldwide sample of wolves. The population of wolves having the most similar, or even identical, sequences compared to the dogs would then be presumed to be the wolf population from which the domestic dogs descend, and the geographical region in which these wolves live would be presumed to be the geographical origin of the dog. However,

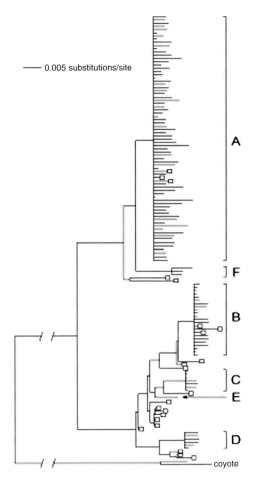

Fig. 2.1. Phylogenetic tree of dog (unlabelled) and wolf (open squares) mtDNA types (Savolainen *et al.*, 2002). Six clades (A–F) of dog mtDNA types are indicated. Branch lengths are according to the indicated scale, the branch leading to the outgroup (coyote) has been reduced by 50%. Clades A–D and F were imploded, and the relationships between mtDNA types within the clades are instead shown as minimum-spanning networks in Figs 2.2 and 2.3.

there is little geographic structure in the mtDNA of wolves, probably because wolves are extremely mobile animals. In a study of the worldwide mtDNA sequence variation among wolves (Vila *et al.*, 1999a) it was found, for example, that a Bulgarian and a Saudi Arabian wolf had identical mtDNA types, and that a Mongolian and another Saudi Arabian wolf shared another type. Thus, the finding of a few wolf mtDNA types closely related to domestic dog types is not in itself a sufficient basis for the determination of the location of domestication. The wolf populations have also experienced severe reductions in population size or total extinction in large parts of their original geographical distribution. There is therefore a possibility that the wolf population from which the dogs evolved has become extinct, rendering such studies impossible. The other approach would be to make a study of today's domestic dog population, making sure to obtain a representative sample from all parts of the world, and comparing the genetic variation between geographic regions, a so-called intraspecific study of the phylogeographic mtDNA variation among dogs. The geographic region having the largest genetic variation would then be assumed to be the region in which the domestic dog originated, based on the assumptions that only part of the genetic variation in the region of origin was spread to the other parts of the world, and that any subsequent mixing of populations through migration would not have been thorough enough to conceal these differences.

Based on the second approach, 654 dogs sampled throughout the world and 38 Eurasian wolves were studied for 582 base pairs of the mtDNA CR (Savolainen et al., 2002). This resulted in the first comprehensive picture of the genetic variation among domestic dogs worldwide. Importantly, it gave a good representation of dogs in Europe, Southwest Asia and East Asia, the three regions that have been the strongest candidates as geographical origin of the domestic dog (Table 2.1). A phylogenetic analysis distributed the mtDNA sequences into the six clades found earlier among Japanese and European breeds, additional samples from other parts of the world thus not adding more major phylogenetic groups in the dog mtDNA phylogenetic tree (Fig. 2.1). The relations between the mtDNA molecules at large genetic distances can be relatively truthfully depicted in the phylogenetic tree, but the data are not sufficient for obtaining a full resolution of dog mtDNA types within the clades. Therefore, the branching in the six dog clades was imploded, and the relations between the mtDNA types within the clades were instead displayed as minimum-spanning networks (Fig. 2.2). The distribution of the dog mtDNA sequences into six clades interspersed by wolf sequences shows that the dog originates from at least six female wolf lines, and possibly many more individual female wolves having identical mtDNA types. Whether this also indicates several different places and time-points for the origin of dogs from the wolf will be discussed below. The six clades were represented by very different numbers of dogs. The largest clade, A, comprised 71% of all dogs, and the three largest clades, A, B and C, a total >95% of the dogs, compared to 3% for clade D and only three dogs each for clades E and F.

The representation of samples from most parts of the world allows a phylogeographic study of the data. Comparing the genetic variation between different parts of the world, the similarities in the distribution of sequences belonging to the three major clades is striking (Table 2.1). All the three major clades A–C are present in all parts of the world except Arctic America, and at similar frequencies. Thus, these three clades, comprising >95% of the dog mtDNA sequences, constitute a common source for a very large proportion of the mtDNA genetic variation in all domestic dog populations. The absence of clades B and C in Arctic America can probably be attributed to a genetic bottleneck in the forming of this dog population; probably this population was

Table 2.1. Number and proportion of individuals, and number of mtDNA types and unique mtDNA types, for the phylogenetic clades A–F and for all six clades (Total) in different populations. The dogs represent a geographical region based on that they either are of a breed of known geographical origin or that they have been sampled in a geographical region with little import of foreign dogs. mtDNA types are defined by substitutions only, disregarding indels.

	Clade A		Clade B		Clade C		Clade D		Clade E		Clade F		Total	
Africa	30 (85.7)	14 (5)	4 (11.0)	1 (0)	1 (2.9)	1 (0)	0	0	0	0	0	0	35	16 (5)
Arctic America	25 (100)	5 (1)	0	0	0	0	0	0	0	0	0	0	25	5 (1)
Europe	140 (67.6)	20 (9)	36 (18.0)	4 (1)	12 (5.9)	3 (0)	19 (9.1)	3 (3)	0	0	0	0	207	30 (13)
East Asia	192 (73.8)	44 (30)	39 (15.0)	10 (7)	24 (9.2)	4 (1)	0	0	3 (1.1)	1 (1)	2 (0.8)	2 (2)	260	61 (41)
SW Asia	51 (56.7)	16 (4)	32 (35.0)	4 (2)	5 (5.4)	2 (0)	2 (2.2)	1 (1)	0	0	0	0	90	23 (7)
Siberia	17 (70.8)	9 (1)	2 (8.3)	1 (0)	4 (17.0)	2 (1)	0	0	0	0	1 (4.1)	1 (1)	24	13 (3)
India	11 (84.6)	4 (0)	1 (7.7)	1 (0)	1 (7.7)	1 (0)	0	0	0	0	0	0	13	6 (0)
Total	466 (71.3)	71	114 (17.4)	13	47 (7.2)	5	21 (3.2)	4	3 (0.5)	1	3 (0.5)	3	654	97

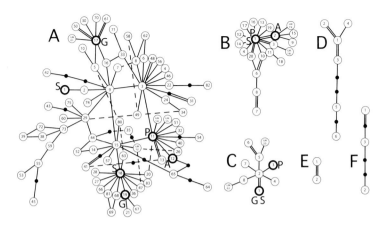

Fig. 2.2. Minimum-spanning networks showing genetic relationships among dog mtDNA types of clades A–F based on 573 base pairs sequence of the mtDNA CR (Angleby and Savolainen, 2005). mtDNA types (white circles) are separated by one mutational step, single strokes represent substitutions and double strokes represent indels. Black dots are hypothetical intermediates. The largely random distribution of breeds and morphological types of dogs among clades A, B and C is exemplified by marking of the mtDNA types found in German Shepherd (mtDNA types A17, A19 and C1, marked 'G'), Pekingese (mtDNA types A11, B1 and C3, marked 'P'), Samoyed (mtDNA types A1, A18, B1 and C1, marked 'S') and Afghan dog (mtDNA types A12 and B2, marked 'A'), and by highlighting of the respective mtDNA types by bold circles.

introduced to America by the ancestors of today's Inuits rather than by the ancestors of the First Nations tribes. Other parts of America were not represented because the original First Nations tribes' dog populations are thought to have been largely obliterated or mixed with European dogs in connection with the arrival of Europeans, which would make the interpretation of American dog sequences complicated. However, in a study of mtDNA in 19 Mexican hairless dogs, a Mexican breed believed to have remained essentially isolated, the mtDNA types all belonged to clades A, B and C (Vila *et al.*, 1999b). Furthermore, the study of archaeological remains of pre-European American domestic dogs (Leonard *et al.*, 2002) showed that among 13 samples from Mexico, Peru and Bolivia, 800 years or more old, 12 had mtDNA types clustering with clade A and one sample had a type clustering with clade B. Thus, both clades A and B, and possibly clade C, according to the data on the Mexican Hairless Dog, were present among the native American dogs. Perhaps more importantly, none of the American samples had sequences which did not fall into the same mtDNA-clades as the Old World samples. Thus, in the available mtDNA data there is no sign of a separate domestication of dogs in America, or of crossbreeding with the wolf. This shows that the dog populations of the New World have a common origin with the dogs of the Old World.

An important observation is that the frequencies of clades A, B and C had similar proportions in all regions (Table 2.1). This suggests that, unless there has been a very effective gene flow along the Eurasian continent, mixing the gene pools along the continent, the major present-day dog populations have had a common origin from a single gene pool containing clades A, B and C. If the three clades had different geographical origins, from separate domestications of the wolf, the frequencies of the clades would be expected to vary between regions. Had, for example, clade A originated in Europe and clade B in East Asia, we would expect to find a higher frequency of clade A in Europe than in East Asia and, vice versa, a highest frequency of clade B in East Asia, with intermediate frequencies in Southwest Asia. The only major deviation from the uniform distribution of proportions of the three clades is found in the Southwest Asian sample, which has a higher frequency of clade B sequences and a lower frequency for clade A sequences than other regions. This could indicate that clade B originated in Southwest Asia separately from clades A and C. However, if this was the case it would also be expected that the genetic variation among the sequences belonging to clade B would be higher in Southwest Asia than in other regions. On the contrary, it was considerably lower in Southwest Asia than in East Asia, and equal comparing Southwest Asia and Europe. This will be further discussed below, but is exemplified by the number of mtDNA types belonging to clade B for the three regions (Table 2.1). A single first origin for the three clades was further indicated by the lack of division of the main morphologic types of dog (spitz, mastiff, greyhound), or of large and small breeds, between the three main clades, except for a lack of greyhounds in clade C (Fig. 2.2). This suggests that the extreme morphologic variation among dog breeds is not the result of different geographically distinct domestications of the wolf. It is impossible today to find out how much the dog populations have been mixed through migration during the thousands of years since the domestication of the wolf, and whether this could explain the universal presence, at similar proportions, of clades A, B and C. However, the similarities over long distances are striking. A comparison of dog breeds from the two island-groups of the British Isles and Japan, situated outside opposite parts of the immense Eurasian continent, offers a good example (British Isles: 81.5%, 13.8% and 4.6%; Japan: 62.5%, 18.8% and 15.6%, for clades A, B and C, respectively). Thus, there are strong indications of a single original gene pool containing the three major dog-mtDNA-clades, A, B and C (Savolainen *et al.*, 2002).

The Geographic Origin of Dogs

An indication of the geographic origin of the three dog clades, and thereby the origin of the dog, can be obtained from a comparison of the genetic variation between geographical regions. If an ancestral population and a derived population (formed from a subset of the genetic types of the ancestral population) are compared, the number of mtDNA types and the nucleotide diversity are expected to be higher in the ancestral population. Comparing Europe, Southwest Asia and East Asia, the three regions that have been the strongest candidates as geographical origin of the domestic dog, East Asia had a larger genetic variation for clades A and B as measured by a number of methods, while for clade C there were no significant differences (Savolainen *et al.*, 2002). Thus, for clade A, the mean pairwise sequence distance, which gives a rough measure of the genetic variation, was 3.39 (SD=0.13) substitutions in East Asia, 2.28 (SD=0.23) in Southwest Asia and 2.97 (SD=0.08) in Europe. Furthermore, the number of mtDNA types, and the number of types unique to the region, were much larger in the East Asian sample than in the European and the Southwest Asian ones (Table 2.1). The difference between the East Asian and the European samples is striking, and, when corrected for sample size by resampling to correct for different sample size, there were significantly more mtDNA types in clade A among 51 East Asian dogs, than among the 51 Southwest Asian dogs, 20.2 compared to 16. Also compared to the other regions there were more mtDNA types in East Asia when corrected for sample size, but not reaching a significant level for the small samples of Africa, Siberia and India. It is notable that out of the 44 mtDNA types found in East Asia, 30 were unique to this region. Thus, the number of mtDNA types unique to East Asia was larger than the total number of types in Europe (Table 2.1). For clade B, the mean pairwise sequence distance was larger for East Asia (0.93 substitutions, SD=0.17) than for Europe (0.45, SD=0.14) and Southwest Asia (0.36, SD=0.11), the East Asian sample had significantly more mtDNA types than those of Europe and Southwest Asia, and a majority of all mtDNA types were unique to East Asia. The difference in number and distribution of mtDNA types in East Asia, Europe and Southwest Asia can be studied in detail in minimum-spanning networks of the three clades (Fig. 2.3). For clade A it is notable that East Asian mtDNA types were distributed throughout the network, while for Europe and Southwest Asia, parts of the network, largely the same in the two populations, were empty.

To conclude, in a comparison of the dog populations of East Asia, Europe and Southwest Asia, there is a much higher genetic variation in East Asia for the two major dog mtDNA-clades, clades A and B, which comprise >90% of all domestic dog mtDNA sequences. This suggests an origin somewhere in East Asia for the mtDNA types of clades A and B, and a subsequent spread of only a subset of these mtDNA types to the rest of the world. For clade C there is no significant difference between the regions, but the similar frequency of clade C among dog populations of the Old World suggests a common origin of clade C together with clades A and B. Thus, the available mtDNA data indicate that the domestic dog has a single geographical origin, somewhere in East Asia.

Dating the Origin

The time of origin for each dog mtDNA clade can be estimated from the mean genetic distance of the sequences in each clade to the original wolf mtDNA type,

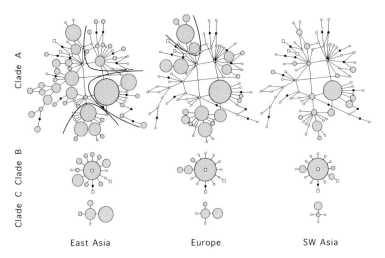

Fig. 2.3. Minimum-spanning networks showing genetic relationships among mtDNA types of phylogenetic clades A, B and C (Savolainen *et al.*, 2002). mtDNA types (circles) are separated by one mutational step, ignoring indels. mtDNA types found in East Asia, Europe and Southwest Asia are indicated in separate networks. The sizes of grey circles are proportional to mtDNA-type frequency in the respective populations. Small uncoloured circles denote mtDNA types not found in the regional population. Black dots are hypothetical intermediates. Uncoloured squares are wolf mtDNA types. Subclusters of clade A discussed in the main text, three in the East Asian and one in the European network, are marked by black lines.

and the mutation rate of the analysed region. Under the assumption that the clade originates from a single wolf mtDNA type, the age of the clade will approximately correspond to the origin of the dogs originating from that female wolf. However, there are two problems with the available data that render a precise dating of the origin of the dog based on the available mtDNA data impossible.

The first problem concerns the dating of the palaeontological fix-point needed to calibrate the mutation rate. The wild canid most closely related to the wolf is the coyote, and the time for the split into two species from their common ancestor is the best available calibrating point for the mutation rate of dog mtDNA. Counting the number of sequence differences between wolf and coyote in the analysed mtDNA region, and dividing by the divergence time, gives the mutation rate. However, while a first appearance of wolves ~700,000 years ago and of coyotes ~1 million years ago (Kurtén, 1968; Kurtén and Anderson, 1980) indicates a date for the divergence between the two species of approximately 1 million years ago, the time of divergence has not been definitely established and an earlier date of up to approximately 2 million years cannot be ruled out. Therefore, without more exact palaeontological evidence, a precise dating of the origin of the dog will not be possible, but it can nevertheless give an indication of the probable range of time, and this information can be used together with other data from the archaeological record on the first appearance of dogs. To simplify the following argument, the 1-million-year date for the wolf–coyote divergence will be used for the calculations based on mtDNA, but bearing in mind the possibility that the split occurred up to 2 million years ago, in which case the age of the datings should be doubled. The second problem is that the mutation rate of the analysed 582 base pairs region does not give resolution between mtDNA types in the time scale needed to monitor the last 20,000 or 30,000 years. According to the calculations above, the rate of mutation is approximately one mutation per 24,000 years in a lineage (Savolainen *et al.*, 2002). This implies that if two wolves, having mtDNA types differing by a single mutation, were domesticated say 15,000 years ago, the dogs of today originating from those two lineages would have mtDNA types differing from each other only by the single original mutation or by just one or two additional mutations. If there were several wolves having mtDNA types differing by just a few mutations, it would not be possible to fully resolve the mtDNA types of the dogs originating from them. Thus, the period around 15,000 years ago, which is suggested to be the time for the origin of the domestic dog according to the archaeological record, cannot be fully studied using the 582 base pairs mtDNA region available at present.

In a domestication event with a subsequent population expansion, a star-like phylogeny, with the founder mtDNA type in the centre and new mtDNA types distributed radially, would be expected. The networks of clades B and C are star-like, indicating an origin from a single wolf mtDNA type (Figs 2.2 and 2.3). In contrast, clade A has a complicated pattern without an easily identifiable central node. A distance of up to 11 substitutional steps between mtDNA types would indicate that clade A is much older than clades B and C, and derives from an initial domestication of wolves. However, as discussed above, the dog mtDNA types in clade A do not necessarily originate from a single wolf mtDNA type even though clade A is an almost completely continuous group of mtDNA types separated from each other by single mutational steps. It is possible that clade A originally was a clade of several closely related wolf mtDNA types, and that several wolves having a number of these mtDNA types belonging to clade A were domesticated. Looking more closely at clade A it has, instead of a single central node, several subclusters with star-like shape, suggesting that clade A may have originated from several wolf mtDNA types (Figs 2.2 and 2.3). The approximate age of clade A, assuming a single origin from the wolf and a subsequent population expansion, is estimated from the mean pairwise distance between East Asian dog-mtDNA-sequences (3.39 substitutions, SD=0.13) and the mutation rate to 41,000 ± 4000 years. This calculation may be biased by the population history among the dogs and wolves. Alternatively, the maximum age of clade A can be estimated from the number of steps between the most distantly related mtDNA types, 11 substitutions apart which corresponds to ~120,000 years. According to these calculations clade A would have originated 40,000–120,000 years ago, and if it is supposed that it was formed in a domestication event from a single wolf mtDNA type, the domestic dog would have originated 40,000–120,000 years ago. If instead an origin of clade A from several different wolf mtDNA types is assumed, several reasonably defined subclusters can be found. To give an alternative dating of the domestication of the dog, three subclusters of clade A possibly representing three origins of dog from wolf, marked by lines in Fig. 2.3, can be studied. The mean genetic distances of the sequences belonging to these subclusters to their respective nodes (0.45, 0.65 and 1.07 substitutions with SD=0.13, 0.09, 0.27, respectively) give estimates of 11,000 ± 4000, 16,000 ± 3000 and 26,000 ± 8000 years for their ages, respectively. Thus, clade A has a substructure, suggesting that it could have been formed by several wolf mtDNA types, possibly ~15,000 years ago. Assuming single wolf mtDNA types as founders of clades B and C, the mean distances among East Asian sequences to the nodes (0.54 and 0.71 substitutions, SD=0.08 and 0.10) give estimated ages of 13,000 ± 3000 and 17,000 ± 3000 years for clades B and C, respectively. Thus, clade A was formed ~40,000–120,000 years ago but has a substructure indicating later population events, and clades B and C were formed ~15,000 years ago. Depending on how these data are interpreted, it can suggest a first origin of domestic dogs either ~40,000–120,000 years ago forming only clade A, after which clades B and C were introduced into the dog gene-pool through crossbreeding between dogs and wolves ~15,000 years ago, but it can also suggest a single origin ~15,000 years ago involving all the three clades A, B and C.

The key question for the dating of the origin of the dog is the number of wolf mtDNA types that formed dog-clade A, but this question cannot be answered based on the present mtDNA data. However, indirect evidence in the form of the universal presence of clades A, B and C at similar proportions talks in favour of a simultaneous origin of the three clades. Assuming that clade A would have originated from an initial domestication approximately 40,000 years ago, the dog population originating from that domestication event would be expected to have spread to other parts of the world. An introduction of clades B and C, 15,000 years ago, into the already existing domestic dog gene pool, by regional crossbreedings with wolves, would require a very thorough mixing to have occurred, through migration to all parts of Eurasia, to level the frequencies of the three clades to the very similar proportions now found throughout the world. Alternatively, if clade A originated from a domestication 40,000 years ago, these dogs could have remained isolated in the original geographical region, not spreading to other parts of the world. Clades B and C would then, 15,000 years ago, have been introduced into the domestic dog gene pool by regional crossbreedings with wolves, after which the dogs would have spread to other parts of the world. While the amount of migration and trade 15,000 years ago is not known, this scenario seems unlikely. It is also contradicted by the age of the oldest subcluster of clade A in Europe (as determined from the mean genetic distance from mtDNA types unique to the western part of the world to the nodal mtDNA type shared with East Asia; 0.39 substitutions, SD=0.09), which is estimated to be only 9000 ± 3000 years old (Fig. 2.3).

Thus, a first origin of dog mtDNA clade A, a long time before the origin of dog clades B and C, does not seem probable considering the very similar proportions of clades A, B and C around the world. The total sum of circumstantial mtDNA evidence therefore indicates a single origin, in both place and time, for the three clades, approximately 15,000 years ago. Furthermore, in the absence of a clear result from the mtDNA data, the best evidence for the time of the first origin of the domestic dog remains the archaeological record, which indicates an origin approximately 15,000 years ago. A synthesis of mtDNA data: the similar proportions of clades A, B and C, and the larger genetic variation for clades A and B; and archaeological data: the oldest evidence of domestic dog dated at approximately 15,000 years ago, therefore points to an origin of the domestic dog in East Asia ~15,000 BP. In this event, clade A would have had several origins from wolf mtDNA types, at least around ten, and the first domestication of wolves would not have been an isolated event, but rather a common practice in the human population in question.

Possible Crossbreedings with Wolves

Clades D, E and F have been left out of the discussion so far to simplify the discussions. These three clades are found only regionally, and clades E and F also at very low frequency, three dogs for each clade, making population genetic analyses virtually impossible (Table 2.1). The three clades, which represent three separate origins from the wolf, could either be the results of crossbreedings between male dogs and female wolves or separate domestications of the wolf, or they could have originated together with clades A, B and C at a single domestication event. Since clades E and F are found in East Asian dogs, an origin together with clades A, B and C seems possible. In this case they would have been part of a gene-pool containing the five clades A, B, C, E and F, but clades E and F, like some parts of clade A, would not have spread to other parts of the world. For clade D the situation is very different. This clade is found only in Europe and Turkey, and it has therefore probably an origin outside East Asia, independent from that of the other clades. There is a clear geographical division within clade D. mtDNA types D1–D4, which constitute a separate subcluster, were all found in Scandinavian spitz breeds, in total 18 out of 49 Scandinavian dogs, while D5 and D6, which are several mutations away from D1–D4, were found in two Turkish and one Spanish dog, respectively (Fig. 2.2). The distance between the subclusters D1–D4 and D5–D6, three substitutions, corresponds to approximately 36,000 years. It therefore seems probable that clade D originates from at least two separate origins from the wolf. The divergence within subcluster D1–D4, a single substitution in 18 individuals, indicates a recent origin, a few thousand years at most. Thus, at least clade D seems to have originated through crossbreeding between male dogs and female wolves. This crossbreeding seems to have had some impact on the Scandinavian breeds, considering the large proportion of dogs having mtDNA types belonging to clade D,

18 out of 49 dogs (37%). It is probable that, in crosses between male dogs and female wolves, the chances of the offspring becoming part of the domestic dog population would be low. The other scenario, that offspring from male wolf/female dog crossbreedings would 'become dogs' seems more probable. It is believed that in some cultures, especially in the northern parts of Eurasia and America, dogs, and above all bitches, were crossed with wolves in order to bring new blood to the dog population. Future studies of the Y chromosome will no doubt tell us to what degree such crossbreedings have contributed to the development of domestic dogs.

The Origin of the Australian Dingo

An interesting question connected to the history of the domestic dog is the origin of the Australian dingo. The dingo is a wild canid, morphologically resembling South Asian domestic dogs, which according to the archaeological record arrived in Australia between 3500 and 12,000 years ago (Clutton-Brock, 1995). However, the precise ancestry and time of arrival in Australia of the dingo have not been known, nor whether, on its arrival, it was domesticated or half-domesticated before becoming feral, or a truly wild dog. In a study of 211 dingoes, sampled around Australia, 582 base pairs of the mtDNA CR were analysed and compared to dog and wolf data, giving very distinct results (Savolainen *et al.*, 2004). All dingo sequences clustered in dog clade A, forming a single internal cluster of mtDNA types unique to dingoes, around a central type, A29, found in both dingoes and dogs (Fig. 2.4). Among domestic dogs, A29 was found only east of the Himalayas, together with most types in that part of clade A. This indicates that the dingo population originates from a single introduction of a small population of domestic dogs coming from East Asia, carrying mtDNA type A29. The mean distance among the dingo sequences to A29 indicates their arrival to Australia ~5000 years ago. The dingo has probably remained isolated since then, and represents a unique isolate of early undifferentiated dogs.

Summary

Studies of physical and behavioural characteristics, and of molecular genetic markers, give consistent evidence that the domestic dog originates from the wolf, and the archaeological record indicates that the wolf was domesticated approximately 15,000 years ago. Population genetic data, based on analysis of mtDNA, shows that all dogs originate from a common gene pool, and indicates that they originated at a single domestication event in East Asia, from which they spread to all continents, including Australia where the domestic dogs developed into the feral dingo. The mtDNA data give only two clear indications of crossbreeding between dog and wolf, indicating that crossbreeding between male dogs and female wolves has contributed only marginally to the dog gene-pool. Studies of Y-

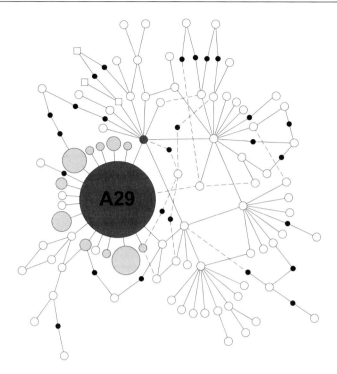

Fig. 2.4. Minimum-spanning network of the main dog clade, A (Savolainen *et al.*, 2004). mtDNA types (circles) and hypothetical intermediates (solid dots) are separated by one mutational step (substitutions, indels not shown). White circles are mtDNA types found in dogs only; light grey circles are types found in dingoes only; dark grey circles are types found in both dingoes and dogs. Squares denote wolves. Areas of grey circles are proportional to frequencies among dingoes, but the area of A29 is reduced by 50%.

chromosomal and autosomal haplotypes, which will no doubt be initiated within a few years, are necessary to test the validity of the conclusions obtained from the mtDNA data, and will, by giving the male side of the story, help to create an even more detailed picture of the evolution of the dog.

References

Angleby, H. and Savolainen, P. (2005) A study of the forensic usefulness of the mitochondrial DNA variation among and within populations, breeds and types of domestic dogs. *Forensic Science International* 154, 99–110.

Clutton-Brock, J. (1995) Origins of the dog: domestication and early history. In: Serpell, J. (ed.) *The Domestic Dog, its Evolution, Behaviour, and Interactions with People.* Cambridge University Press, Cambridge, pp. 7–20.

Clutton-Brock, J. and Noe-Nygaard, N. (1990) New osteological and C-isotope evidence on mesolithic dogs: companions to hunters and fishers at Starr Carr, Seamer Carr and Kongemose. *Journal of Archaeological Science* 17, 643–653.

Houpt, K.A. and Willis, M.B. (2001) Genetics of behaviour. In: Ruvinsky, A. and Sampson J. (eds) *The Genetics of the Dog.* CABI, Wallingford, UK, pp. 371–400.

Kurtén, B. (1968) *Pleistocene Mammals of Europe.* Aldine, Chicago, Illinois.

Kurtén, B. and Anderson, E. (1980) *Pleistocene Mammals of North America.* Columbia University Press, New York.

Leonard, J.A., Wayne, R.K., Wheeler, J., Valadez, R., Guillen, S. and Vila, C. (2002) Ancient DNA evidence for Old World origin of New World dogs. *Science* 298, 1613–1616.

Li, Q. (1990) The history of dog breeding in China. *Chinese Journal of Cynology* 2, 25–27 (in Chinese).

Nobis, G. (1979) Der älteste Haushund lebte vor 14,000 Jahren. *Umschau* 79, 610.

Okumura, N., Ishiguro, N., Nakano, M., Matsui, A. and Sahara, M. (1996) Intra- and interbreed genetic variations of mitochondrial DNA major non-coding regions in Japanese native dog breeds (*Canis familiaris*). *Animal Genetics* 27, 397–405.

Olsen, S.J. (1985) *Origins of the Domestic Dog.* University of Arizona Press, Tucson, Arizona.

Sablin, M.V. and Khlopachev, G.A. (2002) The earliest ice age dogs: evidence from Eliseevichi 1. *Current Anthropology* 43, 795–798.

Savolainen, P., Zhang, Y., Luo, J., Lundeberg, J. and Leitner, T. (2002) Genetic evidence for an East Asian origin of the domestic dog. *Science* 298, 1610–1613.

Savolainen, P., Leitner, T., Wilton, A., Matisoo-Smith, E. and Lundeberg, J. (2004) A detailed picture of the origin of the Australian dingo, obtained from the study of mitochondrial DNA. *Proceedings of the National Academy of Sciences, USA* 101, 12387–12390.

Tchernov, E. and Valla, F.R. (1997) Two new dogs, and other Natufian dogs, from the southern Levant. *Journal of Archaeological Science* 24, 65–95.

Tsuda, K., Kikkawa, Y., Yonekawa, H. and Tanabe, Y. (1997) Extensive interbreeding occurred among multiple matriarchal ancestors during the domestication of dogs: evidence from inter- and intraspecies polymorphisms in the D-loop region of mitochondrial DNA between dogs and wolves. *Genes and Genetic Systems* 2, 229–238.

Vila, C., Savolainen, P., Maldonado, J.E., Amorim, I.R., Rice, J.E., Honeycutt, R.L., Crandall, K.A., Lundeberg, J. and Wayne, R.K. (1997) Multiple and ancient origins of the domestic dog. *Science* 276, 1687–1689.

Vila, C., Amorim, I.R., Leonard, J.A., Posada, D., Castroviejo, J., Petrucci-Fonseca, F., Crandall, K.A., Ellegren, H. and Wayne, R.K. (1999a) Mitochondrial DNA phylogeography and population history of the grey wolf *Canis lupus*. *Molecular Ecology* 8, 2089–2103.

Vila, C., Maldonado, J. and Wayne, R.K. (1999b) Phylogenetic relationships, evolution, and genetic diversity of the domestic dog. *Journal of Heredity* 90, 71–77.

Wayne, R.K. (1993) Molecular evolution of the dog family. *Trends in Genetics* 9, 218–224.

Origin of Dog Breed Diversity 3

C. Vilà and J.A. Leonard

Introduction

Modern dog breeds demonstrate a stunning level of morphological and behavioural diversity. The magnitude of morphological differences between dog breeds is greater than that between all wild species in Canidae, the dog family (Wayne, 1986a, b). This diversity led Charles Darwin and Konrad Lorenz to speculate that perhaps dogs derived from more than one wild ancestral species. However, molecular genetic data has shown conclusively that domestic dogs derive only from the grey wolf (*Canis lupus*) and not any other wild canid (Seal *et al.*, 1970; Sarich, 1977; Ferrell *et al.*, 1978; Wurster-Hill and Centerwall, 1982; Wayne and O'Brien, 1987; Wayne *et al.*, 1987a, b, 1997; see Chapter 2 in this volume). This brings us to the next, obvious question: where did all of the diversity we observe in dogs today come from?

Much has been learnt about the origin of dogs thanks to many years of archaeological research. This research is based on the study of all sorts of material that could indicate prehistoric associations between dogs and humans, from bones to art. During the last 10 years, the development of molecular genetics techniques has opened a new field of research, complementary to the archaeological investigation, which can contribute to a better understanding of the domestication of the dog. In this chapter we describe some molecular techniques and some studies that have employed molecular techniques to clarify some aspect of the process by which dogs were domesticated and modern dog breeds were formed. However, as we will see, these studies have not yet led to a complete understanding of the process and many questions remain to be answered.

Molecular Methods to Study Dog Origins

DNA contains the genetic information that each one of us has inherited from our parents and that codes for most of our physical characteristics. This information is coded by a chain of four nucleotides (normally symbolized as G, A, T and C; for example, a short DNA sequence could be represented: GAAATCTCAGGCA). The total length of the DNA chains (chromosomes) in a typical human cell is about 3 billion nucleotides. All the members of the same species have very similar DNA sequences, which are different from the DNA in other species. The study of the changes observed in the nucleotide sequences can inform us about the evolutionary history of the species. Such changes accumulate over time and may be uniquely shared among taxa. Hence nucleotide changes are markers of common ancestry, and if they accumulate in a regular fashion with time, may be used to estimate divergence time and time of origin. As an example, consider hypothetical sequence changes in the DNA sequence responsible for the synthesis of haemoglobin, the protein responsible for binding oxygen in red blood cells. Suppose all coyotes (*Canis latrans*), grey wolves and dogs shared 90 nucleotides in a fragment of 100 in the haemoglobin gene, whereas the grey wolf and dog shared, in addition, eight more nucleotides. Consequently, the dog and wolf differ by two unique substitutions. Such nested changes allow construction of a phylogenetic tree where each node and branch is defined by a unique set of nucleotide changes (see example in Fig. 3.1). Moreover, if the fossil record suggested that wolves and coyotes diverged about 1 million years ago, and they differ by 10 nucleotide changes, then the two nucleotide differences between dogs and wolves imply a divergence time of 200,000 years ago. Of course, this assumes nucleotide substitutions accumulate in a clock-like fashion with little variation and, although the use of molecular data to estimate divergence time is a common practice, considerable debate surrounds the accuracy of divergence dates (Avise, 1994, 2000).

Many genetic studies of dog domestication are based on the comparison of mitochondrial DNA sequences. Among all DNA, a small part (about 16,000–18,000 nucleotides in length in mammals) is located inside the mitochondria and is inherited just from the mother. This is the mitochondrial DNA (mtDNA). Since this molecule is inherited only from the mother and is not mixed with paternal DNA, it is particularly simple to study. This has made the analysis of mtDNA sequences one of the main tools used to reconstruct the evolutionary relationships between species. The process of domestication needs to be studied with markers that have the highest mutation rates in the mitochondrial genome because divergence between dogs and wolves is extremely recent on an evolutionary time scale. This high mutation rate allows for the relatively fast accumulation of information. This circumstance has led many researchers to focus their investigation on the control region, a fragment of the mitochondrial DNA that might be free of strong selective forces and where most mutations are likely to persist.

Although there are many features of mtDNA that make it a very useful tool, it records only the history of the female lineage. In some cases the history of the

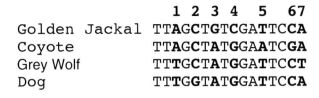

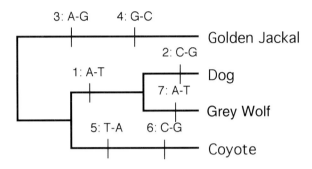

Fig. 3.1. Example of the use of DNA sequence for the construction of phylogenetic trees. The mutations that differentiate the sequences are indicated in bold and identified by numbers. These mutations are used to build the phylogenetic tree, and the position of each mutation is indicated on the corresponding branch in the tree together with the nucleotide change involved. The outgroup (golden jackal in our case) corresponds to a taxon closely related to the group under study. Its sequence is used to give polarity to the tree and provide information on the branching order. In this example, dog and grey wolf differ by two substitutions and each of them differs from coyote by four substitutions.

males and females of a population or breed may be very different. For this reason other markers which have different modes of inheritance complement mtDNA studies very well. For example, the Y chromosome records only the male lineage because it is passed only from father to son, and nuclear autosomal markers are inherited from both parents (Fig. 3.2). Most Y chromosome and nuclear autosomal markers which have been studied so far are microsatellites, simple repetitive sequences which vary in size, and are considered neutral. Some work has also been done on regions of the autosomal genome which are thought to be under selection, such as the major histocompatibility complex (MHC).

Number of Origination Events

Mitochondrial DNA sequences from dogs have been collected to determine the number of origination events in dogs. By themselves these sequences have little value, but the same region was also sequenced in a large number of grey wolves from across their range. The sequences identified in dogs were compared to those

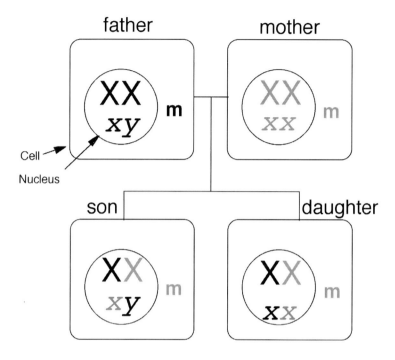

Fig. 3.2. Schematic representation of inheritance of different types of genetic markers in mammals; black represents paternal DNA, grey represents maternal DNA, X represents autosomal chromosomes, x and y represent the sex chromosomes and m represents mitochondrial DNA.

identified in wolves, and a phylogenetic tree was constructed (Fig. 3.3). Most of the dogs formed a single group in the phylogeny, called a clade. This clade, which contains about 80% of the dog sequences, is called clade I (Fig. 3.3; Vilà et al., 1997). In the original study three other minor clades were also identified (Vilà et al., 1997). Since then two more minor clades have been identified (Savolainen et al., 2002). These data could support either a single origination event which gave rise to clade I and a small number of subsequent hybridization events or a small number of origination events (Vilà et al., 1997, 1999; Savolainen et al., 2002). This pattern of a small number of origination events identified by mtDNA sequence surveys characterizes the origin of many livestock mammals (Bruford et al., 2003), for example: sheep (Hiendleder et al., 2002), goats (Luikart et al., 2001), cattle (Troy et al., 2001), pigs (Giuffra et al., 2000), water buffalo (Lau et al., 1998) and donkey (Beja-Pereira et al., 2004). In many cases this view is supported by archaeological studies (Clutton-Brock, 1999). Despite the existence of a limited number of well-defined clades of mtDNA sequences, Savolainen et al. (2002) inferred that dogs derived from a unique centre of domestication. This was due to the concentration of divergent mtDNA sequences in East Asia and the authors assumed that this would imply that dogs had been domesticated there. However, it is not clear that modern dogs can be used to infer patterns of diversity in the past. A study of ancient dog remains has shown that dogs living in America today are genetically very different from the dogs living there 500–1000 years ago (Leonard *et al.*, 2002). The number of domestication events for dogs thus remains unresolved.

Fig. 3.3. Phylogenetic tree of wolf (W) and dog (D) mitochondrial DNA control region sequences (Vilà *et al.*, 1997). Dog haplotypes are grouped in four Clades, I to IV. Boxes indicate haplotypes found in the 19 Mexican hairless dogs (Vilà *et al.*, 1999).

Time of Domestication

Two different approaches have been used to date the time when the domestication of dogs took place. The first one is to try to identify bone remains whose morphology corresponds to dogs and not to wolves (the ancestor of the domestic dog). The second method consists of estimating the date when the mtDNA sequences present in dogs separated from the sequences in wolves. Although both methods target the moment when the dog and wolf lineages split, they look at different things and the dates obtained with both methods are expected to be different. To understand this, imagine one lineage that at a certain time in the past splits in two, as indicated in Fig. 3.4. This could correspond to the split of dogs from the ancestral wolf lineage, and the goal of the studies about domestication is to characterize when this separation occurred. It is at this point that the morphology of the two species starts to diverge. However, it will take some time for their morphology to be sufficiently divergent as to allow the unambiguous identification of bone remains as belonging to one species or the other. Zeder and Hesse (2000) and Zeder (2003) have shown that for other domestic mammals the morphological differentiation from the ancestor did not occur until some time after the initial domestication, when the domesticates expanded to neighbouring areas where they could not mate with wild animals. Similarly, we could expect that the morphological change in dogs may have occurred some time after the domestication

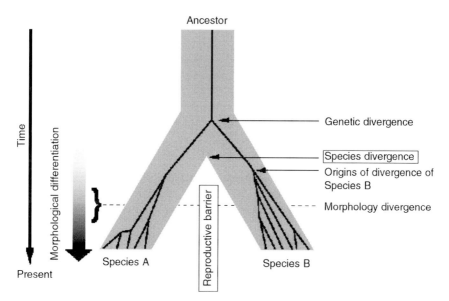

Fig. 3.4. Difference between the species divergence time, and the estimates obtained using genetic and morphological data. Species are indicated as grey bands, and the black lines represent the genetic relationships between sequences from each species. Morphological differentiation between species increases from the time of reproductive isolation.

(although we do not know how much later). Since this is the kind of evidence used by archaeologists, and because the ancient dog bone remains are very scarce, the dates of domestication based on archaeological data will tend to be more recent than the split between dogs and wolves.

Molecular estimates of the time of divergence are based on the comparison of DNA sequences. By applying a molecular clock it is possible to estimate the time necessary to make the two lineages as different as they currently are. In Fig. 3.4, the mtDNA lineages present in wolves and in dogs are represented as the lines inside the grey area. However, the split of the lineages is necessarily older than the split of the populations (the domestication). Consequently, estimates of the date of the domestication based on DNA sequences are likely to be older than the actual event.

Another way to use mtDNA sequences to estimate the time of domestication is to calculate the time of divergence between the most divergent sequences within the domestic clade (the time when dogs started diverging from one another; Fig. 3.4). This estimate is highly dependent on the demographic history of dogs: the origin of these sequences might appear more recent when the population remained small (Avise, 2000). However, it seems logical to think that dogs started from a limited number of individuals and soon after domestication the population started to expand to produce the current population size of several hundred million dogs worldwide (Coppinger and Coppinger, 2001). In this case, the estimates should be reasonably close to the beginning of the dog population expansion, and probably also close to the time of domestication. This is likely a very much better estimator of the age of domestication than the previous ones and should be expected to be younger than the split between dogs and wolves, and older than the first appearance of morphological divergence (see Fig. 3.4).

In summary, while archaeological approaches may lead to an underestimation of the date of domestication, molecular methods tend to produce overestimates. Estimating the age of the diversity within the domestic lineage as opposed to the split between the domestic and wild lineages may produce a more accurate date. While the search for old dog-remains continues, molecular analysis methods are constantly improving, and integrating both approaches will lead us towards a more accurate view of the domestication process.

The archaeological record suggests that dogs were present in the Middle East and Central Europe about 14,000–15,000 years ago (Nobis, 1979; Olsen, 1985; Dayan, 1994; Clutton-Brock, 1999; Sablin and Khlopachev, 2002). Early dogs were morphologically distinct from grey wolves. They are often identified by their smaller body size and wider crania, a more prominent stop on the face and a shortened, crowded jaw (Olsen, 1985; Morey, 1992). The identification of these bones as belonging to dogs and not to wolves implies that the animals had been living with humans for a time period sufficiently long to allow some morphological differentiation from their wild ancestor. The small body size of Asian wolves and the shared presence with early dogs of some traits led to the suggestion that Asian wolves are the direct ancestor of the dog (Olsen and Olsen, 1977).

However, the oldest dog-remains unmistakably identified so far have recently been discovered in western Russia (Sablin and Khlopachev, 2002) and date to 13,000–17,000 radiocarbon years before present (this is equivalent to 16,000–20,000 calendar years before present). These early dogs do not show a reduction in body size when compared to the local populations of grey wolves and have cranial proportions similar to the great dane.

The first attempts to use molecular genetics to study the domestication of dogs took place in the mid-1990s. Okumura et al. (1996) were the first to sequence the control region of a large sample of dogs, but focused on Asian dogs. These authors observed that the mitochondrial DNA sequences grouped into a discrete number of clades. Tsuda et al. (1997) similarly studied a large sample of dogs from Asia, but also included some samples from wolves. Consequently, they were able to convincingly demonstrate a close relationship between wolves and dogs. The close relationship was confirmed in a study published at the same time (Vilà et al., 1997), but in this study an attempt to use the molecular information to estimate the date of domestication was made. The study by Vilà et al. included 162 wolves from 27 populations from throughout Europe, Asia and North America and 140 dogs representing 67 breeds. As Okumura et al. did 1 year earlier, Vilà et al. observed that when constructing an evolutionary tree with all dog sequences, they clustered in just four groups (clades I–IV, Fig. 3.3). One of these groups, clade I, contained most of the studied dogs, and a more recent study including more than 600 modern dogs (Savolainen et al., 2002) concluded that about 71% of them had sequences belonging to this clade. This large diversity could indicate that this group of sequences was the oldest one and thus its origin could correspond to the original domestication event.

To estimate the date of the domestication, Vilà et al. (1997) calculated how much time would be necessary to reach that level of diversity starting from a single mtDNA sequence, using the molecular clock. Vilà et al. first estimated the average sequence divergence between wolves and coyotes for the studied region of the mtDNA and found that it was 7.5%. The fossil record indicates that morphologically differentiated wolves and coyotes existed at least 1 million years ago (Kurtén and Anderson, 1981). Next, the maximum divergence for sequences belonging to clade I was estimated to be 1%. This was used to calculate when the sequences from that clade started to diverge. To confirm that the rate of evolution was not significantly different for the different lineages, the lengths of the different branches in the tree were compared. As no difference was found, the same molecular clock could be used for all the sequences. Finally, since the rate of divergence was around 7.5% per million years, the estimated time needed to generate divergences up to 1% was 135,000 years! The result of the study of Vilà et al. (1997) has been taken as a firm date. However, that estimation was not presented with confidence intervals and these are usually large in this sort of estimate. Also, the region of the mtDNA studied may not be the best to apply a strict molecular clock. On the other hand, the results very clearly suggested that the domestication of dogs had to be much older than indicated by the archaeological record.

More recently, another attempt has been made to use dog mtDNA information to estimate when dogs were domesticated (Savolainen *et al.*, 2002). These authors also focused on the diversity of clade I and concluded that the domestication could have taken place more recently. These authors used a different model of sequence evolution and only considered the sequences in their sample that were currently present on each continent. Since they found the largest diversity in East Asia they concluded that this could have been the place of domestication. Looking at the average sequence divergence among clade I sequences in East Asia, they estimated that the age of the clade was 41,000 years (SD±4000), supporting an origin much older than suggested by the bone remains as the previous study. However, they considered that it was possible that clade I did not derive from one unique domestication event and that several wolf lineages were involved. Clade I was arbitrarily subdivided into various smaller groups and the ages of these sub-clades were estimated to be between 11,000 and 26,000 years old. Despite this wide range of dates, they suggested an East Asian origin for the domestic dog 15,000 years ago (Savolainen *et al.*, 2002). There is a series of problems with this approach. First, East Asia was suggested as the centre of domestication because of its higher diversity. However, whereas most of the samples from the rest of the world corresponded to purebred dogs, this region was characterized by mongrel dogs and dogs of non-recognized breeds. As these animals do not belong to inbred lines it seems likely that they may have a much higher genetic diversity for this reason alone, and this can bias the results. Second, the patterns of current diversity may be very different from ancestral patterns (see the study by Leonard *et al.*, 2002), so looking at modern dogs may not be appropriate to reconstruct patterns of diversity that existed in one region in the past. Third, no objective evidence indicating that clade I derived from one or more founding wolf lineages was provided. Subjectively dividing the clade in smaller groups can produce whatever result the researcher wishes to obtain. Lastly, despite estimating a very old date of origin for one of their sub-clades, 26,000 years (confidence interval: 18,000–34,000 years), the authors conclude that the domestication took place about 15,000 years ago.

Although the results of this study apparently fit with the dates derived from archaeological remains, the conclusion that dogs could have been domesticated in East Asia 15,000 years ago faces one main problem: dog remains (already morphologically differentiated from wolves) of about that time have already been discovered in several places in Europe (Nobis, 1979; Sablin and Khlopachev, 2002) and Asia (Dayan, 1994; Clutton-Brock, 1999), and probably also existed in America (Leonard *et al.*, 2002). The analysis of mtDNA sequences in pre-Columbian dogs of Leonard *et al.* (2002) has shown that American dogs clearly had Eurasian origin, implying that they arrived in the Americas with the first humans. Since all these dogs in Europe, Asia and America shared a common origin, we can presume that the domestication of the dog had to take place significantly before 15,000 years ago to allow their expansion over three continents.

Place of Domestication

As mentioned above, the dog was domesticated from the grey wolf. Therefore, the place of domestication must be somewhere in the range of this species. Grey wolves are naturally distributed across Europe, Asia and North America, so those are the initial possibilities. Very little phylogeographic patterns have been found in grey wolves worldwide, but American and Eurasian animals do not share haplotypes (Vilà *et al.*, 1999) and dog haplotypes are more closely related to Eurasian wolf haplotypes than American wolf haplotypes (Vilà *et al.*, 1997; Leonard *et al.*, 2002). This narrows the range of possibilities to Eurasia. Within Eurasia, divergent lineages have been found in the grey wolves of the Himalayas and Indian subcontinent, which are quite divergent from those found in any dogs (Sharma *et al.*, 2003). This excludes this region as a possible centre of domestication. This narrows the geographic range of domestication to somewhere in Eurasia except India or the Himalayas. Other methods will be required to narrow the area down further.

In order to more precisely identify the place of domestication, Savolainen *et al.* (2002) compared the amount of haplotype diversity in dogs in different regions of the world. They concluded that a single domestication event may have taken place in East Asia. They based their conclusion in the presence of a larger diversity of sequences in this region compared to other regions around the world. However, these conclusions may have been highly influenced by the predominance of dogs that did not belong to pure breeds recognized by the International Dog Federation (FCI) in East Asia. While purebred dogs are characterized by high levels of inbreeding, mongrel dogs show a very high genetic diversity (Irion *et al.*, 2005). This implies that a group of dogs that has not been subjected to the extreme selection process applied to purebred dogs registered in kennel clubs is expected to have a much higher genetic diversity than a similar group of purebred dogs. Additionally, it is not clear to what degree a sample of modern dogs can fully characterize past patterns of diversity. The study of Leonard *et al.* (2002) showed that modern purebred American dogs are very different from the dogs that existed there a few centuries ago. Human history, demography, migrations and even fashion have clearly affected the composition of modern dog populations (Valadez, 1995) and the information about past patterns of diversity are blurred. Consequently, the study of modern dogs may not be the best way to study the place of origin of dogs, and studies of ancient dog remains as well as extensive archaeological research may be needed to provide a final answer to the questions about the number of domestication centres and the place of origin of all modern dogs. However, the answer to this question may not be simple, and recent genetic studies suggest that the separation between wild and domestic lineages may not have been clean cut during part of the dog's history (see next section).

Early Breeding Practices

The first molecular analysis of the origin of dogs used mitochondrial control region sequences to construct a phylogeny of domestic dog and grey wolf haplotypes (Vilà et al., 1997). Four clades of dog sequences were evident in this phylogeny, suggesting that at least four female wolf lineages founded or were subsequently introduced into domestic dogs (Fig. 3.3). A more extensive analysis by Savolainen et al. (2002) suggested six or more lineages. However, the number of actual founding lineages is uncertain because many of the initial matrilines may have been lost through stochastic processes (Avise, 2000) and others may have been introduced through backcrossing between dogs and female grey wolves. The primary conclusion to be derived from these mtDNA data is that a limited number of founding events explain the genetic diversity of dogs. This result is consistent with genetic studies of other domesticated animals (cattle, sheep, goats, pigs; Bruford et al., 2003). By contrast, horses have incorporated numerous matrilines from their wild progenitors, suggesting repeated domestication (or backcrossing) events over a large geographic area (Vilà et al., 2001). However, the limited number of founding events for dogs does not necessarily imply that the number of founders was small.

The major histocompatibility complex (MHC) encodes proteins that are essential to the normal functioning of the immune system and typically show a high level of genetic diversity that is maintained by balancing selection (Hughes and Yeager, 1998). One consequence of this form of selection is that MHC alleles are maintained for longer periods of time than expected under neutrality. Recently, a large number of alleles have been described for the MHC in dogs, wolves and coyotes (Kennedy et al., 1999, 2000, 2001, 2002; Seddon and Ellegren, 2002, 2004). Seddon and Ellegren (2002) observed that a large number of alleles are shared between the three different species (Fig. 3.5), which suggests that they may have existed before the species diverged (over 1 million years ago for grey wolves and coyotes). Hence, these alleles can provide evolutionary information over a longer time frame than mtDNA and have been used to provide estimates of effective population sizes in humans (Ayala et al., 1994) and to calculate founding population sizes for other species (Vincek et al., 1997). Predictive models based on selection and drift can be used to estimate the minimum numbers of wolf founders under a variety of demographic conditions (Vilà et al., 2005).

Vilà et al. (2005) collated information from 42 MHC *DRB* alleles for dogs, almost all of them differing by at least one non-synonymous change. Considering the rate of non-synonymous substitution estimated for MHC class II loci (Klein et al., 1993), the authors concluded that almost the totality of the alleles probably originated before the time of domestication. Consequently, at least 21 founders are required to explain the large MHC diversity observed in dogs. However, this number represents a minimum estimate since it assumes that all founders are heterozygous for different alleles, are equally successful producing offspring and that no alleles are removed from the population by drift. To obtain more realistic

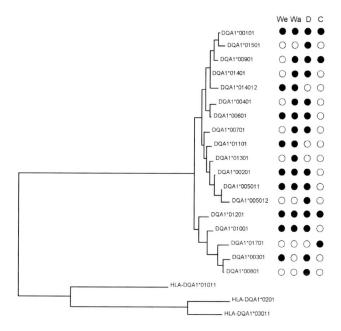

Fig. 3.5. Phylogenetic tree of *MHC DQA* alleles. The species distribution of each allele is indicated by filled circles; We, European grey wolf; Wa, North American grey wolf; D, dog; C, coyote. Many alleles are shared between species, indicating that the alleles may have originated before the species split. Based on Seddon and Ellegren (2002).

estimates of the number of founders, Vilà *et al.* (2005) used models that varied the number of founders, the number of founding populations, the rate of population growth after domestication, the diversity in the ancestral population and the strength of selection acting on the MHC locus. These simulations strongly suggested that if the domestication occurred in only one population, hundreds of individuals must have been involved in the founding of the domestic dog (Fig. 3.6). Obviously, this is not a realistic possibility at the time of the domestication of the dog because humans were still hunter-gatherers, living in small groups, and they lacked the resources to support a large population of domestic animals. Since several lineages of mtDNA have been identified in dogs (Vilà *et al.*, 1997, 1999; Savolainen *et al.*, 2002), one possibility was that the domestication would have taken place in different populations. If this was the case, drift would lead to the fixation of different alleles in each population, and a lower number of founders would be enough to explain the level of diversity found today. Indeed, the simulations show that the estimated number of founders decreased as the number of populations involved increased (Fig. 3.6). However, even if six populations were involved, which is more than suggested by mtDNA studies (Vilà *et al.*, 1997, 1999; Savolainen *et al.*, 2002) and more than suggested for most livestock species

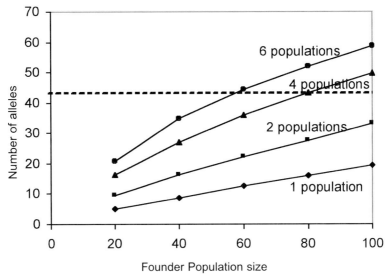

Fig. 3.6. Number of MHC alleles that would be expected in modern dogs depending on the number of wolves involved in the founding event (divided in 1–6 populations of equal size). Each dot represents the average of 1000 simulations. The discontinuous line indicates the number of alleles described so far in modern dogs. To obtain such a large number of alleles, a large number of founders would be required, even in the case that dogs derived from multiple domestication events. Modified from Vilà et al. (2005).

(Bruford et al., 2003), about 60 founders would be required. This is an underestimate considering that the model assumes that all individuals are equally likely to reproduce and, consequently, the real number of founders would probably be several times larger (Frankham, 1995).

In conclusion, the simulations showed that the number of wolves involved in the domestication process was larger than deduced from mtDNA alone and probably involved several hundred wolves. Second, since it seems improbable that such a large founding population could have been kept separated from wild wolves, hybridization between wolves and dogs was probably an important influence on the diversity of the dog MHC. Presumably, extensive backcrossing with male wolves early in the history of dog domestication may have enriched nuclear diversity to a greater extent than maternally inherited mitochondrial DNA. Future research utilizing male-specific markers on the Y chromosome may help clarify the situation.

Origin of Breeds

The World Canine Organization (Fédération Cynologique Internationale, FCI) currently recognizes about 347 breeds of dogs, classified in ten groups according to their function and, to a lesser degree, area of origin. Each one of these breeds is characterized by a quite unique morphology and often also behaviour, which makes them easily recognizable. The genetic basis of these morphological traits is evident since breeds breed true (i.e. the offspring of two dalmatians looks like another dalmatian, and the offspring of a couple of boxers will look like another member of the same breed). Similarly, some breed-characteristic behaviours are also inherited (Schmutz and Schmutz, 1998). This uniform typology for each breed is very often seen as the result of a long period of isolation. In fact, many dog breeders believe that their breeds had an origin that dates from several centuries or millennia ago (Crowley and Adelman, 1998). For example, archaeological evidence from ancient Egypt suggests that several types of morphologically differentiated dogs similar to mastiffs and greyhounds existed there 4000 years ago and Romans may have been the first people to develop dog breeds in Europe (Clutton-Brock, 1999). Also, pottery from the Colima culture (250 BC–AD 450, western Mexico) clearly represents Mexican hairless dogs (Cordy-Collins, 1994). This breed was also found by the Spanish conquistadors on their arrival in Mexico in the 16th century (Valadez, 1995). Furthermore, dogs depicted in European paintings during that time could be easily recognized today as spaniels, mastiffs, hounds, pointers, etc.

Consistent with this differentiation at the phenotypic level, multiple studies using autosomal markers (located on nuclear chromosomes other than the sex chromosomes X and Y) have shown that breeds are clearly differentiated (Lingaas et al., 1996; Zajc et al., 1997; Zajc and Sampson, 1999), and that the genotype obtained for an individual using only a limited number of markers may be sufficient to assign that individual to the correct breed (Koskinen, 2003; Parker et al., 2004; Halverson and Basten, 2005). Parker et al. (2004) typed 96 microsatellite loci in 85 breeds, each represented by five dogs. An analysis of molecular variance showed that variation between breeds accounted for about 27% of the genetic diversity observed in all dogs. Similarly, they also assessed genetic diversity within and between breeds with nuclear single nucleotide polymorphisms (SNPs). In a survey of 19,867 base pairs of sequence from 120 dogs, the authors identified 75 SNPs, and 14 of these were breed-specific in the small number of individuals screened per breed. These results suggest many breeds have been isolated since their origination.

The clear differentiation of breeds in phenotype and with autosomal genetic markers stands in stark contrast to the maternally inherited mtDNA results. Studies based on mitochondrial DNA found no correlation between mtDNA haplotype and breed (Vilà et al., 1997, 1999; Savolainen et al., 2002). A single breed often had many different mtDNA haplotypes derived from different sequence clades. For example, Vilà et al. (1999) screened 19 Mexican hairless dogs and found seven different mtDNA haplotypes that fall into three of the four main clades of dogs (Fig. 3.3). Also, a single mtDNA haplotype could be found across a wide variety of breeds. For example, haplotype D3 from Vilà et al. (1997) was identified in chow chow, Norwegian elkhound, Mexican hairless, Siberian husky, papillon, poodle, rottweiler, English setter, Icelandic sheepdog, springer spaniel,

Japanese spitz, border terrier, fox terrier and whippet. This lack of differentiation between breeds for mitochondrial DNA markers is surprising given the morphological uniformity within breeds compared to the large differences between breeds and suggests that modern breeds have a recent origin from a well mixed and genetically diverse dog population.

In order to reconcile the lack of differentiation between breeds at mtDNA with the clear morphologic separation, Sundqvist et al. (2006) studied paternally inherited Y chromosome markers, maternal mtDNA and autosomal markers in purebred dogs. The authors observed, again, that breeds were genetically differentiated for the autosomal markers while both mtDNA and Y chromosome were not well separated. However, the patterns of diversity observed for these two markers were highly informative about the process by which the breeds were formed. When looking at the number of different Y chromosome and mtDNA types found in five dogs from each of 20 breeds, Sundqvist et al. observed that the number of mtDNA haplotypes was usually larger than the number of Y chromosome haplotypes (Table 3.1). Since these two marker systems evolve in a very different way, the results could be different to interpret. However, the authors also typed the same markers in several populations of wolves (Table 3.1), where approximately the same number of males and females reproduce every year (Packard, 2003) and observed similar numbers of mtDNA and Y chromosome types. The pattern observed in the wolf populations could be used as a baseline to understand the results in dog breeds, and the authors concluded that the different pattern was due to the process by which dog breeds were formed. Fewer males than females had contributed to the origin of the breeds. Since males have a larger breeding potential than females (one male can sire a large number of litters every year), the results suggest that breeders have traditionally selected traits present in males to define the breed characteristics. This has resulted in the existence of 'popular sires' (Ostrander and Kruglyak, 2000), dogs that, because of their characteristics, have often been used as breeders. Some of these males may produce over 100 litters in their life time (Ostrander and Kruglyak, 2000). Consequently, a phenotypic trait present in one male can be successfully transferred to a larger number of offspring per generation than if that trait was in a female. Therefore, strong selection on males allows a faster definition of the phenotype in a new breed.

Although the comparison of the levels of diversity for uniparentally inherited markers within dog breeds was informative about the process by which the breeds were formed, they were unable to provide meaningful information about the relationship between dog breeds. Attempts to reconstruct phylogenetic relationships between breeds have been made using autosomal markers, but simple relationships could not be identified when a large number of breeds were considered (Irion et al., 2003; Parker et al., 2004). Parker et al. (2004) found that many of the most recognizable and common breeds of European origin are so closely related that branching patterns were difficult or impossible to elucidate. This suggests a very recent origin for all of these breeds, and also that they share a common recent ancestry from a diverse population of dogs. This is consistent with the

Table 3.1. Number of mtDNA and Y chromosome haplotypes observed in dog breeds (5 males per breed) and in wolf populations (*n*=number of males sampled per population). Based on Sundqvist *et al.* (2006).

Dog breed/wolf population	mtDNA	Y chromosome
Dogs		
Airedale terrier	3	2
Beagle	3	1
Bernese mountain dog	2	1
Border terrier	2	1
Boxer	3	1
Cairn terrier	3	1
Cavalier King Charles spaniel	2	1
Collie, rough/smooth	3	1
Dalmatian	3	2
Flatcoated retriever	2	1
German pointer	3	1
German shepherd	3	1
Golden retriever	2	1
Greyhound	3	2
Irish soft-coated wheaten terrier	4	2
Newfoundland	1	1
Poodle, miniature/standard	4	2
Rottweiler	1	2
Shetland sheepdog	3	3
West Highland white terrier	3	2
Wolves		
Alaska (*n*=12)	6	6
Russia (*n*=12)	4	6
Inuvik (*n*=13)	4	7
Finland (*n*=31)	3	7
Spain (*n*=20)	2	5
Baltic States (*n*=24)	2	9

foundation of dog breeding clubs, breed standards and modern breeding practices (Ash, 1927; Crowley and Adelman, 1998; Dennis-Bryan and Clutton-Brock, 1988; Parker *et al.*, 2004).

The World Canine Organization (FCI) has organized breeds into ten different groups based on function, and, to some extent, supposed origin. However, the genetic analysis of Parker *et al.* (2004) defined only four fundamental genetic subdivisions in dogs. The most primitive and divergent groupings included a wide diversity of breeds ranging in geographic origin from Africa (basenji) to Asia (chow-chow and akita) to the Arctic (Siberian husky and Alaskan malamute). The wide geographic area represented by these primitive breeds suggests that dogs spread over a large geographic area soon after domestication. Three other breed clusters can be defined by the genetic analysis: (i) European breeds that diverged in the 1800s; (ii) herding dogs, including the collie and sheep dogs and similar to FCI group 1 (sheep dogs and cattle dogs); and (iii) mastiff-like dogs, including the bulldog and boxer. The first one of these groups (primitive dogs) may appear so differentiated from all other breeds because the other groups include most of the breeds of European origin and so may have a very narrow genetic basis. On the other hand, the other groups may indicate the preferential exchange of dogs among the breeds included within the groups. Sundqvist *et al.* (2006) have observed that the breeds included in these groups (as well as in the FCI groups) tend to have more similar Y chromosomes than mtDNA sequences, which could indicate preferential exchange of males than females between the breeds. This would be consistent with the use of males to define the characteristics of the breeds and would indicate that the similarity between breeds at autosomal microsatellites may not so much indicate a recent common origin but the occasional exchange of individuals.

An understanding of the origin and evolution of modern breeds is important to models of genetic disease and theories about the genetic basis of morphological and behavioural variation (Wayne and Ostrander, 1999). However, our current knowledge about the origin of dogs and about the process by which breeds were formed which led to the astonishing diversity of modern-day dogs is far from complete. The completion of a second dog genome sequence during 2005 will certainly promote our understanding of the diversity of dogs and we should expect many answers and new questions to arise during the next few years.

Acknowledgements

Financial support was provided by a grant of the Swedish Research Council for Environment, Agricultural Sciences and Spatial Planning to C.V. J.A.L. is supported by a grant from the Swedish Research Council.

References

Ash, E.C. (1927) *Dogs: Their History and Development.* E. Benn Limited, London.
Avise, J.C. (1994) *Molecular Markers, Natural History and Evolution.* Chapman and Hall, New York.
Avise, J.C. (2000) *Phylogeography: The History and Formation of Apecies.* Harvard University Press, Cambridge, Massachusetts.
Ayala, F.J., Escalante, A., O'Huigin, C. and Klein, J. (1994) Molecular genetics of speciation and human origins. *Proceedings of the National Academy of Sciences, USA* 91, 6787–6794.
Beja-Pereira, A., England, P.R., Ferrand, N., Jordan, S., Bakhiet, A.O., Abdalla, M.A., Mashkour, M., Jordana, J., Taberlet, P. and Luikart, G. (2004) African origins of the domestic donkey. *Science* 304, 1781.

Bruford, M.W., Bradley, D.G. and Luikart, G. (2003) DNA markers reveal the complexity of livestock domestication. *Nature Reviews Genetics* 4, 900–910.

Clutton-Brock, J. (1999) *A Natural History of Domesticated Animals*. Cambridge University Press, Cambridge.

Coppinger, R. and Coppinger, L. (2001) *Dogs*. The University of Chicago Press, Chicago, Illinois.

Cordy-Collins, A. (1994) An unshaggy dog history. *Natural History* 2, 34–40.

Crowley, J. and Adelman, B. (1998) *The Complete Dog Book: Official Publication of the American Kennel Club*, 19th edn. Howell Book House, New York.

Dayan, T. (1994) Early domesticated dogs of the Near East. *Journal of Archaeological Science* 21, 633–640.

Dennis-Bryan, K. and Clutton-Brock, J. (1988) *Dogs of the Last Hundred Years at the British Museum (Natural History)*. British Museum (Natural History), London.

Ferrell, R.E., Morizot, D.C., Horn, J. and Carley, C.J. (1978) Biochemical markers in species endangered by introgression: the red wolf. *Biochemical Genetics* 18, 39–49.

Frankham, R. (1995) Effective population size/adult population size ratios in wildlife: a review. *Genetical Research* 66, 95–107.

Giuffra, E., Kijas, J.M., Amarger, V., Carlborg, O., Jeon, J.T. and Andersson, L. (2000) The origin of the domestic pig: independent domestication and subsequent introgression. *Genetics* 154, 1785–1791.

Halverson, J. and Basten, C. (2005) A PCR multiplex and database for forensic DNA identification of dogs. *Journal of Forensic Science* 50, 352–363.

Hiendleder, S., Kaupe, B., Wassmuth, R. and Janke, A. (2002) Molecular analysis of wild and domestic sheep questions current nomenclature and provides evidence for domestication from two different subspecies. *Proceedings of the Royal Society of London, Series B* 269, 893–904.

Hughes, A.L. and Yeager, M. (1998) Natural selection at major histocompatibility complex loci of vertebrates. *Annual Reviews Genetics* 32, 415–435.

Irion, D.N., Schaffer, A.L., Famula, T.R., Eggleston, M.L., Hughes, S.S. and Pedersen, N.C. (2003) Analysis of genetic variation in 28 dog breed populations with 100 microsatellite markers. *Journal of Heredity* 94, 81–87.

Irion, D.N., Schaffer, A.L., Grant, S., Wilton, A.N. and Pedersen, N.C. (2005) Genetic variation analysis of the Bali street dog using microsatellites. *BMC Genetics* 6, 1–6.

Kennedy, L.J., Carter, S.D., Barnes, A., Bell, S., Bennett, D., Ollier, B. and Thomson, W. (1999) DLA-DRB1 polymorphisms in dogs defined by sequence-specific oligonucleotide probes (SSOP). *Tissue Antigens* 53, 184–189.

Kennedy, L.J., Altet, L., Angles, J.M., Barnes, A., Carter, S.D., Francino, O., Gerlach, J.A., Happ, G.M., Ollier, W.E., Polvi, A., Thomson, W. and Wagner, J.L. (2000) Nomenclature for factors of the dog major histocompatibility system (DLA), 1998: first report of the ISAG DLA Nomenclature Committee. *Animal Genetics* 31, 52–61.

Kennedy, L.J., Angles, J.M., Barnes, A., Carter, S.D., Francino, O., Gerlach, J.A., Happ, G.M., Ollier, W.E., Thomson, W. and Wagner, J.L. (2001) Nomenclature for factors of the dog major histocompatibility system (DLA), 2000: second report of the ISAG DLA Nomenclature Committee. *Animal Genetics* 32, 193–199.

Kennedy, L.J., Barnes, A., Happ, G.M., Quinnell, R.J., Bennett, D., Angles, J.M., Day, M.J., Carmichael, N., Innes, J.F., Isherwood, D., Carter, S.D., Thomson, W. and Ollier, W.E. (2002) Extensive interbreed, but minimal intrabreed, variation of DLA class II alleles and haplotypes in dogs. *Tissue Antigens* 59, 194–204.

Klein, J., Satta, Y., O'hUigin, C. and Takahata, N. (1993) The molecular descent of the major histocompatibility complex. *Annual Review of Immunology* 11, 269–295.

Koskinen, M.T. (2003) Individual assignment using microsatellite DNA reveals unambiguous breed identification in the domestic dog. *Animal Genetics* 34, 297–301.

Kurtén, B. and Anderson, E. (1981) *Pleistocene Mammals of North America*. Columbia University Press, New York.

Lau, C.H., Drinkwater, R.D., Yusoff, K., Tan, S.G., Hetzel, D.J. and Barker, J.S. (1998) Genetic diversity of Asian water buffalo (*Bubalus bubalis*): mitochondrial D-loop and cytochrome b sequence variation. *Animal Genetics* 29, 253–264.

Leonard, J.A., Wayne, R.K., Wheeler, J., Valadez, R., Guillen, S. and Vilà, C. (2002) Ancient DNA evidence for Old World origin of New World dogs. *Science* 298, 1613–1616.

Lingaas, F., Aarskaug, T., Sorensen, A., Moe, L. and Sundgreen, P.-E. (1996) Estimates of genetic variation in dogs based on microsatellite polymorphism. *Animal Genetics* 27, 29.

Luikart, G., Gielly, L., Excoffier, L., Vigne, J.D., Bouvet, J. and Taberlet, P. (2001) Multiple maternal origins and weak phylogeographic structure in domestic goats. *Proceedings of the National Academy of Sciences, USA* 98, 5927–5932.

Morey, D.F. (1992) Size, shape, and development in the evolution of the domestic dog. *Journal of Archaeological Science* 19, 181–204.

Nobis, G. (1979) Der älteste Haushund lebte vor 14,000 Jahren. *Umschau* 19, 610.

Okumura, N., Ishiguro, N., Nakano, M., Matsui, A. and Sahara, M. (1996) Intra- and interbreed genetic variations of mitochondrial DNA major non-coding regions in Japanese native dog breeds (*Canis familiaris*). *Animal Genetics* 27, 397–405.

Olsen, S.J. (1985) *Origins of the domestic dog*. University of Arizona Press, Tucson, Arizona.

Olsen, S.J. and Olsen, J.W. (1977) The Chinese wolf ancestor of New World dogs. *Science* 197, 533–535.

Ostrander, E.A. and Kruglyak, L. (2000) Unleashing the canine genome. *Genome Research* 10, 1271–1274.

Packard, J.M. (2003) Wolf behavior: reproductive, social and intelligent. In: Mech, L.D. and Boitani, L. (eds) *Wolves. Behavior, Ecology, and Conservation*. University of Chicago Press, Chicago, Illinois, pp. 35–65.

Parker, H.G., Kim, L.V., Sutter, N.B., Carlson, S., Lorentzen, T.D., Malek, T.B., Johnson, G.S., DeFrance, H.B., Ostrander, E.A. and Kruglyak, L. (2004) Genetic structure of the purebred domestic dog. *Science* 304, 1160–1166.

Sablin, M.V. and Khlopachev, G.A. (2002) The earliest Ice Age dogs: evidence from Eliseevichi. *Current Anthropology* 43, 795–799.

Sarich, V.M. (1977) Albumin phylogenetics. In: Rosenoer, V.M., Oratz, M. and Rothschild, M.A. (eds) *Albumin Structure, Function and Uses*. Pergamon Press, New York, pp. 85–111.

Savolainen, P., Zhang, Y.P., Luo, J., Lundeberg, J. and Leitner, T. (2002) Genetic evidence for an East Asian origin of domestic dogs. *Science* 298, 1610–1613.
Schmutz, S.M. and Schmutz, J.K. (1998) Heritability estimates of behaviors associated with hunting in dogs. *Journal of Heredity* 89, 233–237.
Seal, U.S., Phillips, N.I. and Erickson, A.W. (1970) Carnivora systematics: immunological relationships of bear albumins. *Comparative Biochemistry and Physiology* 32, 33–48.
Seddon, J.M. and Ellegren, H. (2002) MHC class II genes in European wolves: a comparison with dogs. *Immunogenetics* 54, 490–500.
Seddon, J.M. and Ellegren, H. (2004) A temporal analysis shows major histocompatibility complex loci in the Scandinavian wolf population are consistent with neutral evolution. *Proceedings of the Royal Society of London, series B* 271, 2283–2291.
Sharma, D.K., Maldonado, J.E., Jhala, Y.V. and Fleischer, R.C. (2003) Ancient wolf lineages in India. *Proceedings of the Royal Society of London, series B* 271, S1–4.
Sundqvist, A.-K., Björnerfeldt, S., Leonard, J.A., Hailer, F., Hedhammar, Å., Ellegren, H. and Vilà, C. (2006) Unequal contribution of sexes in the origin of dog breeds. *Genetics* 172, 1121–1128.
Troy, C.S., MacHugh, D.E., Bailey, J.F., Magee, D.A., Loftus, R.T., Cunningham, P., Chamberlain, A.T., Sykes, B.C. and Bradley, D.G. (2001) Genetic evidence for Near-Eastern origins of European cattle. *Nature* 410, 1088–1091.
Tsuda, K., Kikkawa, Y., Yonekawa, H. and Tanabe, Y. (1997) Extensive interbreeding occurred among multiple matriarchal ancestors during the domestication of dogs: evidence from inter- and intraspecies polymorphisms in the D-loop region of mitochondrial DNA between dogs and wolves. *Genes and Genetic Systems* 72, 229–238.
Valadez, R. (1995) *El perro mexicano*. Instituto de Investigaciones Antropológicas, Universidad Nacional Autónoma de México, Mexico.
Vilà, C., Savolainen, P., Maldonado, J.E., Amorim, I.R., Rice, J.E., Honeycutt, R.L., Crandall, K.A., Lundeberg, J. and Wayne, R.K. (1997) Multiple and ancient origins of the domestic dog. *Science* 276, 1687–1689.
Vilà, C., Maldonado, J. and Wayne, R.K. (1999) Phylogenetic relationships, evolution and genetic diversity of the domestic dog. *Journal of Heredity* 90, 71–77.
Vilà, C., Leonard, J.A., Götherström, A., Marklund, S., Sandberg, K., Lidén, K., Wayne, R.K. and Ellegren, H. (2001) Widespread origins of domestic horse lineages. *Science* 291, 474–477.
Vilà, C., Seddon, J. and Ellegren, H. (2005) Genes of domestic mammals augmented by backcrossing with wild ancestors. *Trends in Genetics* 21, 214–218.
Vincek, V., O'Huigin, C., Satta, Y., Takahata, Y., Boag, P.T., Grant, P.R., Grant, B.R. and Klein, J. (1997) How large was the founding population of Darwin's finches? *Proceedings of the Royal Society of London, series B* 264, 111–118.
Wayne, R.K. (1986a) Cranial morphology of domestic and wild canids: the influence of development on morphological change. *Evolution* 4, 243–261.
Wayne, R.K. (1986b) Limb morphology of domestic and wild canids: the influence of development on morphologic change. *Journal of Morphology* 187, 301–319.
Wayne, R.K. and O'Brien, S.J. (1987) Allozyme divergence within the Canidae. *Systematic Zoology* 36, 339–355.

Wayne, R.K. and Ostrander, E.A. (1999) Origin, genetic diversity, and genome structure of the domestic dog. *Bioessays* 21, 247–257.

Wayne, R.K., Nash, W.G. and O'Brien, S.J. (1987a) Chromosomal evolution of the Canidae: I. Species with high diploid numbers. *Cytogenetics and Cell Genetics* 44, 123–133.

Wayne, R.K., Nash, W.G. and O'Brien, S.J. (1987b) Chromosomal evolution of the Canidae: II. Divergence from the primitive carnivore karyotype. *Cytogenetics and Cell Genetics* 44, 134–141.

Wayne, R.K., Benveniste, R.E. and O'Brien, S.J. (1989) Molecular and biochemical evolution of the Carnivora. Molecular and biochemical evolution of the Carnivora. In: Gittleman, J.L. (ed.) *Carnivore Behavior, Ecology and Evolution*. Cornell University Press, Ithaca, New York, pp. 465–494.

Wayne, R.K., Geffen, E., Girman, D.J., Koepfli, K.P., Lau, L.M. and Marshall, C.R. (1997) Molecular systematics of the Canidae. *Systematic Biology* 46, 622–653.

Wurster-Hill, D.H. and Centerwall, W.R. (1982) The interrelationships of chromosome banding patterns in canids, mustelids, hyena, and felids. *Cytogenetics and Cell Genetics* 34, 178–192.

Zajc, I. and Sampson, J. (1999) Utility of canine microsatellites in revealing the relationships of pure bred dogs. *Journal of Heredity* 90, 104–107.

Zajc, I., Mellersh, C.S. and Sampson, J. (1997) Variability of canine microsatellites within and between different dog breeds. *Mammalian Genome* 8, 182–185.

Zeder, M. (2003) Hiding in plain sight: the value of museum collections in the study of the origins of animal domestication. In: Grupe, G. and Peters, J. (eds) *Decyphering Ancient Bones: the Research Potential of Bioarchaeological Collections*. Rahden/Westf., Leidorf, pp. 125–138.

Zeder, M.A. and Hesse, B. (2000) The initial domestication of goats (*Capra hircus*) in the Zagros mountains 10,000 years ago. *Science* 287, 2254–2257.

II Biology and Behaviour of Dogs

Introduction

In this part of the book, some fundamental aspects of the behavioural biology of dogs are covered. Chapter 4 deals with the ethological foundation for animal behaviour. The way any animal behaves is a function of its neural system and the way the muscles are controlled by this, both aspects being under heavy evolutionary selection. To understand this evolutionary aspect of behaviour properly, we need to consider how genes may affect behaviour, and this is the topic of Chapter 5. Here we get a broad overview of behaviour genetics including some of the most powerful methods available today to dissect the mechanisms allowing particular genes to affect specific behaviour.

Furthermore, to really understand the behaviour of dogs, we need to be able to perceive the world through their senses. To help with this, Chapter 6 provides an overview of different sensory organs of mammals, their construction and function, with a special emphasis on dogs. These senses help dogs orienting, hunting, etc., but are of course also crucial in their social interactions. Dogs are profoundly social animals, and Chapter 7 puts this aspect of dog behaviour into context by providing extensive comparisons with other closely related canids.

The last chapter of this section, Chapter 8, is devoted to learning. Being able to modify and consistently affect the behaviour of dogs is a prerequisite for many aspects of life with dogs today. This is therefore one of the longest chapters of the book, and it contains a comprehensive overview of modern learning theory, applied to everyday examples of dog learning. The chapter therefore provides an excellent bridge to the third part of the book, which is concerned with dogs among humans.

4 Mechanisms and Function in Dog Behaviour

Per Jensen

Introduction

Ethology is the science in which we study animal behaviour, its causation and its biological function, and this entire book is devoted to the behavioural biology of dogs. But it is not entirely clear what we understand by behaviour in the first place, and some examples may help to reflect on this problem of definition of the subject. In its simplest form, behaviour may be series of muscle contractions, perhaps performed in clear response to a specific stimulus, such as in the case of a reflex – for example, a dog scratching itself with the hindleg. However, in the other extreme, we find very complex activities, such as a pack of wolves seizing a prey, continuously assessing their directions and positions with the help of various cues from the environment, from the behaviour of the prey and from the rest of the pack.

We would use the word behaviour for both of these extremes, and for many other activities in between in complexity. It will include all types of activities that animals engage in, such as locomotion, grooming, reproduction, caring for young, communication, etc. Behaviour may involve one individual reacting to a stimulus or a physiological change, but may also involve two individuals, each responding to the activities of the other. This makes the science of ethology a complex one, and there is a need to structure the way in which to approach the huge task of analysing the biology of behaviour.

Since behaviour is such a complex biological phenomenon, it follows that it can be studied from a number of different perspectives. Any science dealing with behaviour must therefore limit its purpose, or formalize the framework of the science. The most influential formulation of a research programme for ethology was put forward by the Nobel Prize winner Niko Tinbergen (Tinbergen, 1963).

This programme is frequently referred to as 'Tinbergen's four questions', and it outlines the specific aspects that he felt were ethology's central task to study. Assume that we are interested in studying the behaviour of barking in dogs (see also Chapter 12 in this volume for a more thorough discussion of barking). What sorts of questions should we pose? According to Tinbergen, ethology can ask four different types of questions, which are not mutually exclusive. A complete ethological understanding of barking requires that we provide answers on all of them, albeit not necessarily in one and the same study:

1. What causes barking? The answer to this question refers to the immediate causes, such as which stimuli elicit or stimulate a particular behaviour. In this question, we may also ask which physiological variables, such as hormones, cause or mediate the performance of barking.
2. What is the function of barking? In this case, the answer describes how the performance of a behaviour affects the reproductive success, the fitness, of the dog. It therefore has to do with evolutionary aspects and consequences.
3. How does barking develop during ontogeny? Studies of this question aim at describing the way a behaviour is modified by individual experiences. It will also cover questions such as how barking matures, and how it changes with the age of the dog, and how it is affected by learning.
4. How has barking developed during phylogeny? This is a clearly evolutionary question, and usually calls for comparative studies of related species. It is of course not possible to examine the vocal behaviour of fossils, but comparing vocalizations of different now living, related canids, may help to understand when the behaviour developed and how it has been preserved in different lineages of species.

The interested dog biologist – who is likely to have a close personal relation with dogs – might object that Tinbergen's questions are incomplete, since they do not consider what animals perceive, feel and know in relation to their own behaviour, i.e. the emotional and cognitive aspects of behaviour. This aspect of animal behaviour was long considered to be inaccessible to science. However, over the last decades, scientists have developed methods and concepts to allow investigation into this area. This has led to a new branch of the science, emerging in the 1970s, known as cognitive ethology (Bekoff, 2000). In this book, Chapter 12 is largely devoted to these aspects of dog behaviour. However, proper understanding of dog behavioural biology requires that the Tinbergen questions are examined systematically and, based on the answers to them, we can proceed and pose new questions regarding cognitive aspects of behaviour.

Nature and Nurture

An important question concerns the relative contribution of inherited traits (genetic factors) and acquired traits (for example, learned) in the expression of a particular behaviour. This is sometimes referred to as the nature versus nurture debate. Chapter 5 in this book is particularly concerned with the genetic aspects of dog behaviour. Here, I will just emphasize that all behaviour, whether we refer to it as innate or learned, is necessarily always a product of both genetic and environmental factors, i.e. it is always a question of nature AND nurture. This double necessity is particularly evident in some developmental processes, such as the maturation and socialization of dog pups, which is treated later in this chapter.

We can easily understand the wider aspects of the 'nature-and-nurture view' by considering a typical learned behaviour, such as retrieving game. In order for the dog to learn to retrieve on command, it needs some basic behavioural disposition to chase and carry objects. This disposition can be selected for, as we know from specialized retrieving breeds such as labradors, so obviously there is a genetic basis for this behaviour. However, we can teach the dog to connect a signal (for example, a spoken command) with the performance of the behaviour, which is an acquired process – no dog is born with the knowledge that 'fetch' means 'run and fetch that dead bird and bring it to me'. In order to be able to form this connection, the dog needs sensory organs that can perceive the signal (ears in this case), a nervous system that can process it and form the connection between the correct behavioural response and a reward, and a locomotory system (muscles and joints) that allows it to retrieve. All these features are constrained by genetic factors. Therefore, any behaviour will always be the result of an interaction between genetic predispositions and environmental variables.

Behaviour is a Brain-product

Imagine a dog performing a typical canine behaviour: it carries a bone away from the family, digs a hole in the soil, drops the bone in the hole and covers it with soil, using the nose to push the soil in place. It is self-evident that the behaviour of the dog is controlled by the brain, but how? Information reaches the brain from the sense organs (smell and shape of the object, visual information from the surroundings) and is interpreted there in some meaningful way to the dog ('the stimuli represent a bone'; 'this is a suitable place to dig'). Following this interpretation, the brain sends nerve impulses to the appropriate muscles, which make the dog move and start digging. From watching many dogs performing this type of behaviour, one can easily conclude that the pattern is similar for different members of the species: holes are normally dug with specific movements with the forelegs, the bone is normally always covered with the help of the nose. The reason is that the complex of movements is guided by specific programmes by which the brain controls motor activity. If we want to understand the mechanisms of dog behaviour, it therefore seems reasonable to start exploring the brain and the nervous system.

The Brain and the Nervous System

In all vertebrates, dogs included, the brain develops in the embryo from the so-called neural tube (Alcock, 2001). Early, it differentiates into three different parts, the forebrain, midbrain and hindbrain. Somewhat later, the forebrain differentiates further into the telencephalon and the diencephalon, the hindbrain differentiates into metencephalon and myelencephalon, and the midbrain stays as a functionally intact unit in the adult (the mesencephalon). Of course, the relative specialization of the different brain parts varies between species. In mammals, telencephalon develops into cerebrum, which includes the cerebral cortex, and this structure is extremely variable between species. It is the part of the brain containing centres for integration of sensory stimuli, and for conscious reasoning and reflection – in particular a part of the cortex called the neocortex, which is unique to mammals. Humans have the most developed cerebral cortex, and the neocortex makes up about 80% of our total brain mass. Dogs, like other mammals, have the same structures, but they constitute a relatively much smaller part of the brain.

One part of the cerebral cortex is common to all mammals (and reptiles), and is referred to as the limbic system (Alcock, 2001). This consists of the hippocampus and the olfactory cortex (where olfactory input is integrated, see Chapter 6), and parts of the thalamus and hypothalamus of the diencephalon. This region of the brain contains centres involved in the immediate control of basic behavioural programmes, related to feeding, aggression and sexual behaviour. The limbic system also connects to sensory areas in the neocortex and is responsible for attaching emotions, feelings, to behaviours. Human patients receiving electrostimulation in different nuclei of the limbic system report sudden feelings of fear or anger, and so on, depending on the exact localization of the stimulation (Carlson, 1981).

Whereas dogs differ considerably from humans in the structure and size of the cerebral cortex and neocortex, the limbic system is very similar both in structure and relative size. This is, of course, a strong neurobiological argument for the assumption that dogs are able to perceive more or less the same range of feelings as we do, but have a limited ability to reflect consciously on these feelings.

Neurons and their Organization

The information which the brain receives is carried via neurons, specialized cells which are able to propagate an electric current, called an action potential (Carlson, 1981). Neurons connect to each other via a so-called synapse. A synapse may either be excitatory, meaning that the action potential in the first neuron elicits a similar action potential in the second, or inhibitory, in which case the action potential of the first neuron prevents the second from releasing any electrical activity. The central nervous system (CNS) consists of the brain and the spinal cord, both of which are made up of billions of interconnected neurons, which in turn are organized into different nuclei and areas specialized in handling certain types of information processing.

The part of the nervous system not included in the brain and spinal cord is called the peripheral nervous system (PNS). This is the system which receives information from sensory organs of different kinds. Some of them monitor events outside the body, and these are treated in more detail in Chapter 6. Others monitor internal events in the body, such as blood glucose levels, gut extension and muscle tension, sends this information to the brain and allows the CNS to consistently maintain a scheme of the state of the body and its surroundings.

The PNS also has outgoing, efferent, neurons, which innervate muscles and other organs. This division of PNS is divided into two separate systems, the somatic and the autonomic systems. The somatic system controls the aspects of motor activity which humans can consciously affect, for example the skeletal muscles. The autonomic system is responsible for the control of smooth and cardiac muscles, and the gastrointestinal and endocrine systems, and can usually not be controlled consciously. The autonomic system is further divided into the sympathetic and the parasympathetic branch, which together determine the degree of arousal and tension of the body. Neurons from both branches innervate most organs of the body, and the relative strength of activation of each of the branches determines the effect on the organ. For example, the sympathetic neurons increase heart rate, inhibit digestion and increase breathing; parasympathetic neurons do the opposite (Fig. 4.1).

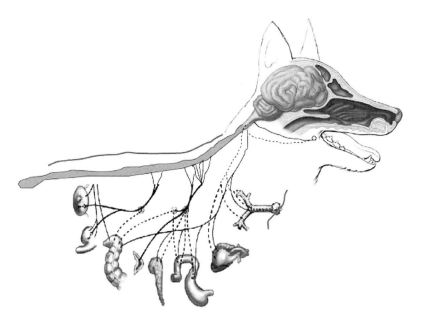

Fig. 4.1. An overview of the innervation of different organs by sympathetic (solid lines) and parasympathetic (dotted lines) parts of the autonomic nervous system. A particular organ is normally innervated by both sympathetic and parasympathetic neurons, which exert opposite effects – if one stimulates the activity of the organ, the other inhibits. In this picture only a few examples of innervated organs are shown: salivary glands, bronchs, heart, stomach, pancreas, adrenal medulla, intestines, sexual organs and bladder.

The protrusions of neurons (axons) from brain and spinal cord are collected in bundles, nerves, which are often embedded in myelin, a protein which protects the neurons and increases transmission speed of action potentials. Some neuron bundles are not myelinated, and have a slower transmission rate.

Stimuli

Stimuli from the surroundings constantly flow over every individual. A small part of all available information is recorded by specialized sense organs, and is transmitted to the brain in the form of visual, olfactory, auditory and mechanic stimuli. The fraction of the available information that is detected and perceived by an animal is first limited by the structure and function of the sensory organs – this is dealt with extensively in Chapter 6.

When the stimuli reach the brain, they are interpreted in a meaningful way to the animal. The male dog brain may identify a collection of olfactory stimuli as, for example, a female in heat, a potential prey, or the territorial mark of an intruding male. Interestingly, the interpretative patterns are to a large extent innate and do not have to be learned. This is the basis of so-called key stimuli – a collection of stimuli is automatically interpreted in a specific, functional way. A male three-spined stickleback (a small fish) reacts to anything roundish with a red ventral part as if it was an intruder – the key stimulus for a rival is simply 'roundish and red at the bottom'. Similarly, dogs do not need specific training to identify the meaning of social signals such as a wagging tail. On the other hand, it takes extensive experience before a dog can learn that a cat wagging its tail is not in a friendly mood, since this violates the innate interpretation pattern of this signal.

Key stimuli are usually linked to specific behavioural responses. A small, furry object which moves away from the dog very quickly will release chasing behaviour. Racing dogs, such as whippets, are fooled into running fast by the key stimuli of a prey attached to an arm on an engine. A dog can learn not to react to a key stimulus, but if not specifically trained, they will behave according to the innate interpretation of the sensory input.

The Motor Apparatus

Motor activity is the central aspect of all behaviour. It occurs as a result of synchronized pathways of muscle contractions, which cause the animal to move in a functional manner in relation to a particular stimulus. The simplest forms of coordinated muscle activity are commonly referred to as reflexes, and these were discovered and studied extensively in dogs by the Russian physiologist Pavlov (see Chapter 8 for further discussion of Pavlov's findings). A reflex is a simple, nonvariable, motor response to a specific stimulus (Toates, 2002). Usually, the reflex involves neurons that do not necessarily connect to the CNS, so

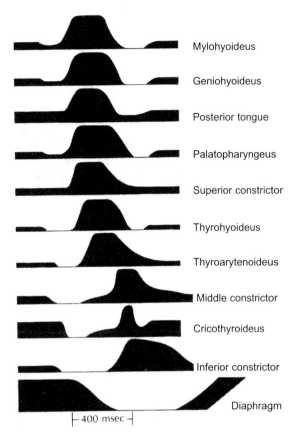

Fig. 4.2. The sequence of contractions of a number of muscles, highly coordinated in order to form the swallowing reflex of the dog. Reprinted with permission from McGraw Hill.

the reflex motion can be carried out without any conscious awareness or feelings. In humans, the knee-jerk reflex is a common example, and in dogs, the swallowing reflex has been extensively studied (Fig. 4.2). Although simple, the reflex may involve a large number of muscles which contract and relax in a highly coordinated and patterned fashion, as can be seen in the swallowing reflex of dogs.

A somewhat more complicated pattern of muscular activity is the so-called modal action pattern (MAP), often referred to as the fixed action pattern. MAPs are again highly structured, but usually more complex than reflexes. They normally involve the CNS, but once elicited, the full motor programme proceeds with minimal interference from the CNS. This means that the MAP usually proceeds to the end once it has been released. For example, a dog will often attempt to roll and rub in substances with a strong smell. Once the behaviour has been released, it will proceed largely in the same manner every time; first the dog bends one foreleg, then it places the shoulder part against the odorous substance and finally rolls over on its back while wriggling the body. The neural programme controlling the behaviour is tightly wired, which makes it virtually impossible for the dog to perform it in a different way (for example, a dog normally can not decide to rub its belly in the smell instead). However, there is some flexibility built into the control system, for example allowing the dog to vary duration and intensity, or to interrupt the behaviour if strongly stimulated – that is why the pattern is preferably called 'modal' rather than 'fixed' (Toates, 2002).

The total motor output of a dog, or any animal, is to a large extent a complex of more or less flexible combinations of series of reflexes and MAPs. Taken together, this can be referred to as 'motor programmes', a term which implies that there is certain level of neural organization of the order and structure of the muscle contractions, but also that there is some flexibility to allow the animal to adjust the motor output to the prevailing circumstances.

In summary, this means that what an animal can do is highly constrained by the organization of the neural system and the muscles. This is one important reason for why animals have clearly identifiable species-specific behaviour. The dog which we met earlier will have great problems in deciding to cover its bone with soil in any other manner than by using its nose – a cat would rather use its forelegs to scratch soil over something it attempts to cover.

Endocrine Moderation of Behaviour

As we have seen, the brain interprets incoming stimuli and controls motor programmes. A particular set of stimuli (key stimuli) tends, to give rise to a particular motor programme. However, this system is controlled in all parts by the chemical environment of the body. Hormones moderate and affect behavioural output in a number of ways, and they are all part of the body's endocrine system (Toates, 2002).

Hormones of various types are secreted from glands in different parts of the body, and they have a variety of biological functions. Some control metabolic pathways and gastro-intestinal activity, others affect sex organs and yet others are linked to the function of the immune system. Most hormones serve many different purposes, but in this chapter we are mostly concerned about those hormones that affect behaviour in various ways.

Some of the most important hormones in this respect are those which are controlled by, or secreted directly by, the hypophysis (pituitary). This gland is located just under the brain, with nerve projections directly from the hypothalamus. It secretes hormones which have a stimulatory effect on other glands in the body, and thereby has a central regulatory role in the endocrine system. For example, the hypophyseal hormone FSH (follicle-stimulating hormone) stimulates the testes of male dogs and causes them to increase testosterone secretion. Testosterone has a broad behavioural effect on the dog; for example, it increases its sexual responsiveness and lowers its threshold for aggression (Toates, 2002).

Another hypophyseal hormone, oxytocin, exerts a direct behavioural effect by calming the dog (Uvnas Moberg, 1994).

Hormones do not cause or inhibit behaviour by themselves. Rather, they affect the sensitivity of the neural pathways involved in different behaviours. For example, testosterone sensitizes the responsiveness for stimuli which are related both to aggression and sexual behaviour. In the male dog, less olfactory stimuli from females in heat, and less visual stimuli from a rival are needed to start the sexual and aggressive motor programmes when testosterone levels are high. Furthermore, testosterone strengthens the motor output, so sexual and aggressive behaviour is performed at a higher intensity when the hormone levels are higher.

Although the endocrine system is dynamic and can change dramatically over a short time, some effects are more lasting than others. For example, testosterone and oestrogen have several important functions in modulating the nervous system during development. A male embryo which is exposed to high levels of oestrogen in the uterus will become feminized in a number of ways (vom Saal *et al.*, 1983). For example, it will generally be less aggressive when adult. This modulation of the nervous system goes on at least up to puberty, and this is the reason why aggressive male dogs castrated late in life may not respond as the owner wishes. When the dog approaches puberty, the nervous system is to some extent already 'calibrated' on a certain aggression threshold, and the daily fluctuations in testosterone levels become less important for the level of aggressiveness.

Development of Behaviour

A newborn dog is a rather helpless creature – the eyes are closed, acoustic abilities utterly limited, and the olfactory system very immature. The motor programmes are limited to a few essential movement patterns necessary to obtain food and get rid of metabolic waste products. However, within a few weeks, the dog has developed a wide array of perception abilities and motor skills, and a few weeks later it has finished some of the most important developmental pathways in its life. This developmental process is a beautiful example of how genetics and experience are both necessary parts of normal behavioural development. The process is often divided into four periods: the neonatal period, the transition period, the socialization period and the juvenile period (Braastad and Bakken, 2002).

During the neonatal period, the pups are completely dependent on their mother. They react to tactile stimuli and possibly smell. Movement is undeveloped and consists of slow crawling and head oscillations. They also emit whining or yelping vocalizations.

During the transition period, the adult behaviour patterns start to develop. The eye-opening at approximately 13 days of age marks the start of the period, which ends when the ear channels open at 18–20 days. During the transition period, the motor abilities develop, and eventually the pup will start walking. Eliminative behaviour no longer requires tactile stimulation from the mother. Play-fighting and tail-wagging also start to develop during this period, and the vocalizations become more variable.

During the period of socialization between 3 and 8 weeks, the young dog starts to display most adult behaviour patterns. During this period, the first signs of fear can also be observed. Social behaviour develops, as can be seen by the tendency to coordinate activities, and the first signs of aggression show up. During this phase, social bonds with other pups and with the mother and humans develop. The most sensitive period in this respect is 4–8 weeks of age. If the pups are deprived of human contact during this period, they may be extremely difficult to tame later on. During this limited time, the nervous system is predisposed for learning what a familiar flock and family member looks and smells like, and this capacity is largely lost after the socialization period.

By 8 months of age, the dog is more or less fully grown. During the juvenile phase, which follows after the socialization period, basic behaviour is relatively invariant, although motor abilities are improved.

Between 6 and 14 months of age, dogs of most breeds enter sexual maturation. Sexual as well as aggressive behaviour develops rapidly. The male dog starts performing raised leg urinations (RLU) and regardless of sex, social rank disputes are common.

Development and Calibration of the Behavioural System

The processes going on during pup development can be formalized with the help of a developmental model suggested by the Canadian scientist Jerry Hogan (Hogan, 1988). Following the reasoning from the previous parts of this chapter, the behaviour system can be said to be made up of three different parts: the perceptional system, the motor system and the central system (Fig. 4.3).

The reaction norms of some parts of each of these systems are genetically predetermined, whereas others require environmental input, learning. For example, the newborn dog probably has no specific knowledge of the concept of 'mother', this must be learnt. But the pathways for which stimuli that should be combined into a 'mother'-concept are predetermined, so the pup will specifically be sensitive for certain combinations of sensations. As soon as it opens it's eyes it will start looking for stimuli complexes that make sense and eventually this complex will be assigned the concept of 'mother'. Note that the genetic predispositions steer in considerable detail what and when to learn – it is therefore impossible to call the developmental process either genetic or acquired – it is both at the same time.

In the same way, some motor output is pre-wired at birth, whereas other output depends on training and acquisition of skills. This ensures that the pattern which develops is both species-specific and adapted to the precise circumstances under which the individual grows up.

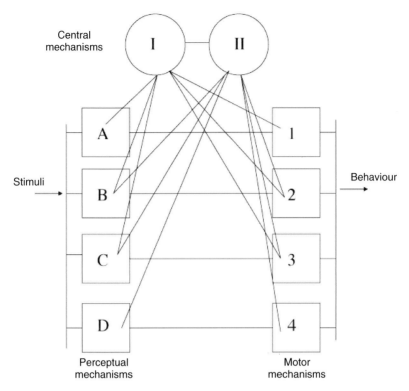

Fig. 4.3. A conceptual model for a behaviour system. Stimuli from the external world are analysed by perceptual mechanisms, which are integrated by central mechanisms or relayed directly to motor mechanisms, the output of which equals behaviour. Central mechanisms may affect several behaviour outputs and be related to several types of sensory input. Redrawn after Hogan (1988).

Motivation and Conflict

Now that we have examined how a particular behaviour develops and is controlled, what remains to be investigated is the central process whereby a dog determines whether or not a particular stimulus is salient enough to warrant action. This will, of course, also depend on which other stimuli impinge on the dog at the moment, and which other activities it is occupied by. Simply spoken, the dog needs to prioritize its activities.

The processes involved in making these kinds of decision are usually referred to as motivation (Toates, 2002). The concept can be loosely defined as the likelihood that an animal will perform a particular behaviour at a certain time. For example, the probability that a dog will start drinking in a certain time period could be taken as a measure of its drinking motivation, and the probability that it will ingest food, of its feeding motivation. Often, motivation is used with a wider meaning to include all processes and rules that determine an animal's decision of whether to engage in a particular behaviour or not.

If we could watch a dog in a situation where only stimuli relevant for one particular behaviour system is present, we might be able to measure its motivation for that behaviour rather precisely – it would be an inverse function of the necessary strength of stimuli needed for releasing a response. For example, imagine that a dog is kept in a closed, evenly lit room without windows. A bowl of food is placed in front of the dog, and we want to measure how long it takes before it starts feeding as a measure of its feeding motivation. If we had been able to keep all other stimuli than food constant, we might then achieve a good measurement.

However, we are not likely to succeed. Even in this highly controlled situation, the dog is exposed to a series of conflicting motivations. First of all, it is alone, so the motivation to gain social contact with other dogs or humans might be high. Second, we cannot stop physiological processes, so the dog may be dehydrated and have a high motivation to drink. Similarly it may be motivated to explore the room, to escape the situation or to test its social status in relation to the person providing the food bowl. So, even in this highly artificial situation, the dog's behaviour is a result of the solution of a conflict between many different competing motivational systems.

Motivations are thought to be neurally arranged in a hierarchical fashion (Fig. 4.4). On the top level, the central motivational decisions are made, such as whether to be active or to sleep. Once a high-level decision has been made, the lower levels follow in a more or less automatic fashion. When the dog has decided to follow its motivation to feed, other motivations are closed down, and the feeding system has the exclusive control over the motor system until some other motivation takes over.

Whereas the strength of some motivational systems are mainly a function of the prevailing sensory input, others fluctuate independent of stimuli. These motivations form the basis of rhythmical behaviour. For example, sleep motivation follows a diurnal rhythm – even in a room with light all day, a dog will settle into a regular diurnal sleep–wake rhythm (Toates, 2002). Some motivational systems are to a certain extent self-regulated – the motor system will be activated at regular intervals, or at a certain time since it was last active (Hogan, 1988). This has the consequence that animals will have an urge to perform some behaviour patterns regularly even in the absence of relevant stimuli. In dogs, chasing and biting may be partly regulated in this way. Behaviours with this type of motivational control are sometimes referred to as ethological or behavioural needs, and the failure to allow for these needs may cause development of abnormal behaviours of different kinds (Jensen and Toates, 1993) (see also Chapter 13).

Stress and Stereotypies

Biological stress is one of the most well investigated physiological and behavioural phenomena, and it has a high significance for dog welfare. Stress can be defined

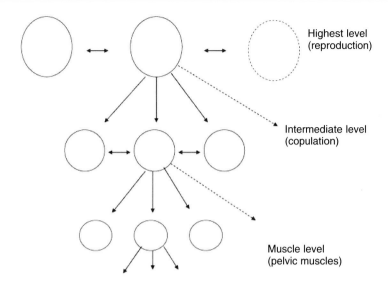

Fig. 4.4. Tinbergen's view of the hierarchical organization of behaviour, in this example applied to the reproductive behaviour of dogs. Behaviour is thought to be controlled on different hierarchical levels, and within each level, the systems are mutually inhibitive. If, for example, reproductive behaviour is in control at the top level, other systems, such as sleep, feeding etc., will be inhibited. The same applies on all levels, so the reproductive system controls the use of the individual muscles, for example to allow the dog to copulate.

as the state of an animal when it is not able to act in accordance with its motivational state (Jensen and Toates, 1997). In this situation, there will normally be elevated levels of cortisol, a hormone secreted by the adrenal cortex following stimulation by the hypophysis hormone ACTH. There will also be an increased activity in the sympathetic nervous system, which is associated with increased heart rate and blood pressure, plus a number of other sympathetically regulated states.

Dogs suffering prolonged periods of high stress will run the risk of developing various diseases. For example, the immune system may be down-regulated, increasing the susceptibility for infections, and the cardiovascular system may be damaged. When a stress situation is chronic or persistent, dogs, like other animals, will try alternative ways to cope with the stress situations, and they may develop various behaviour disorders. A more comprehensive description of behaviour disorders in dogs, and their causes, can be found in Chapter 13.

A typical class of stress-induced behaviour disorders is stereotypies, defined as movements which are repeated over and over again, without any obvious function (Lawrence and Rushen, 1993). These abnormal behaviours are common in zoos, on farms and in laboratory units, where animals often have limited possibilities to express some highly motivated behaviours. They are also not uncommon in dogs living under conditions which do not allow a full expression of normal behaviour. Typically, an animal with a highly motivated, but thwarted, behaviour, will attempt to perform parts of the behavioural programme, and then repeat this part for an extended period of time. Horses with limited locomotory possibilities may develop weaving (swaying from side to side while putting the weight on alternative front legs), and horses with limited possibilities for foraging behaviour may develop crib biting (biting edges of the crib, stall or fence). Common stereotypies in dogs are, for example, tail chasing and wall jumping. Stereotypies may be difficult to cure once they have developed, so to avoid them, preventive measures are most efficient.

Stress can also lead to increased aggression, which the dog may direct either towards other dogs, humans, including the owner, or both (see Chapter 13). The mechanism here may be that the increased sympathetic nerve activity puts the dog in a state of 'fight or flight', which lowers the threshold for aggression-inducing stimuli. However, increased aggression as such is not diagnostic for stress, since many other conditions may give the same result (see Chapters 13 and 14).

Final Remarks

The behaviour of dogs is – just as for any animal – a result of how the nervous system interprets sensory information and transforms this into muscle activity. Sensory organs, the structure of the nervous system, and the ways in which muscles and bones build up the body are all processes that are under the control of genetic factors. However, as we have seen, both the development of the dog, and the processing of information, depend on environmental input. Behaviour is therefore truly an effect of both genes and environment. Understanding the way in which behaviour is controlled and executed is essential for understanding how dogs experience their lives and why some individuals sometimes develop abnormal and unwanted behaviour.

References

Alcock, J. (2001) *Animal Behavior – An Evolutionary Approach*, 7th edn. Sinauer Associates Inc., Sunderland, Massachusetts.
Bekoff, M. (2000) Animal emotions: exploring passionate natures. *Bioscience* 50, 861–870.
Braastad, B. and Bakken, M. (2002) Behaviour of dogs and cats. In: Jensen, P. (ed.) *The Ethology of Domestic Animals – An Introductory Text*. CABI, Wallingford, UK, pp. 173–193.
Carlson, N.R. (1981) *Physiology of Behavior*. Allyn and Bacon, Toronto, Ontario.
Hogan, J.A. (1988) Cause and function in the development of behavior systems. In: Blass, E.M. (ed.) *Handbook of Behavioral Neurobiology*. Plenum Press, New York, pp. 3–15.

Jensen, P. and Toates, F.M. (1993) Who needs 'behavioural needs'? Motivational aspects of the needs of animals. *Applied Animal Behaviour Science* 37, 161–181.

Jensen, P. and Toates, F.M. (1997) Stress as a state of motivational systems. *Applied Animal Behaviour Science* 54, 235–243.

Lawrence, A.B. and Rushen, J. (eds) (1993) *Stereotypic Animal Behaviour – Fundamentals and Applications to Welfare*. CAB International, Wallingford, UK.

Tinbergen, N. (1963) On aims and methods of ethology. *Zeitschrift für Tierpsychologie* 20, 410–433.

Toates, F.M. (2002) Physiology, motivation and the organization of behaviour. In: Jensen, P. (ed.) *The Ethology of Domestic Animals – An Introductory Text*. CAB International, Wallingford, UK, pp. 31–50.

Uvnas Moberg, K. (1994) Role of efferent and afferent vagal nerve activity during reproduction: integrating function of oxytocin on metabolism and behaviour. *Psychoneuroendocrinology* 19, 687–695.

vom Saal, F.S., Grant, W.M., McMullen, C.W. and Laves, K.S. (1983) High fetal estrogen concentrations: correlation with increased adult sexual activity and decreased aggression in male mice. *Science* 220, 1306–1309.

5 Behaviour Genetics in Canids

Elena Jazin

Introduction

Behavioural genetics is a relatively new field joining scientists interested in the study of the modifications and variation of behaviours due to genetic differences. Genes are the units of heredity passed down from parents to offspring, and they are not enough, by themselves, to finally determine all behaviours. Instead, they operate together with environmental conditions. Environment has a different meaning in behavioural genetics than in other contexts. As an ecological term, environment means the physical world. As a genetic word, environment means all influences other than inherited factors, including external influences such as weather conditions or interactions with other organisms, but also factors inside the body such as nutrients, hormones or pathogens, to cite a few of multiple possibilities. In general all behaviours are due to biological processes in the brain, and the state of the brain is influenced by multiple factors that shape the products of the function of the brain, which are thoughts and behaviours. Behavioural genetics is a multidisciplinary field that overlaps and complements research related to biology, psychology, physiology, medical genetics, evolutionary science and neuroscience among others. To clarify a common confusion, it should be mentioned that this research area is not called behavioural genetics because all scientists have a reductionistic view of the problem, and think that genes are more important than environment to determine behaviour. Instead, the name is due to the fact that these scientists use genetic-based research tools to dissect all factors, genetic and environmental, that contribute to differences in behaviour.

Are All Behaviours of Interest for Behavioural Genetics?

One of the reasons why this field faces a very difficult task is that behaviour is not always easy to define and measure with validity and reliability. Behaviour, in a very general sense, can be defined as the actions a living organism makes, as a whole, in response to the environment (see also Chapter 4 in this volume). Anything from a sneeze due to pollen, crawling under a tree to avoid the sun, barking or courtship before mating, are behaviours. Physical manifestations of most diseases, such as epileptic seizures, can be considered behaviours as well. Although slightly expanded in relation to the view proposed in Chapter 4, we can broadly say that there are two forms of behaviour, those that are actions and those that are states of mind, not necessarily externally manifested. Therefore the term behaviour can also include mental processes such as thought, dreams or emotions, such as depression, aggression or happiness. Some behaviour is unique to one individual, for example the way your dog puts his pad on the door twice when he wants to be let outside. Other behaviours are just unique to one situation a certain individual is exposed to, and they may not be repeated, such as the way you would react if you suddenly see an astronaut in your backyard. Other behaviours are uniform across a whole species or breed. Consider for example the instinctive retrieval behaviour of a yellow labrador or the herding posture of a border collie. The last kind of behaviours, usually called instincts or behavioural adaptations, are most interesting for behavioural genetics. We can say that a behavioural adaptation is an inborn pattern of activity, or tendency to action, common to a given biological species or breed. It is a performance by an animal of complex acts, absolutely without instruction or previously acquired knowledge. In some schools of research, for many years the only instinct commonly accepted was a general ability to learn, but today we recognize many different ones, ranging from internal manifestations such as fears and phobias, to complex combination of physical acts, such as mating strategies. Behavioural adaptations are inherited from one generation to the next and, therefore, they are influenced by gene modifications. However, even those behaviours that have a genetic component can be modified by environmental influences, for example a dog's innate ability to retrieve can be nurtured, shaped and trained to different levels.

What is the Molecular Basis of an Inherited Behaviour?

Genes are the basic unit of heredity and are located on chromosomes, condensed strands of genetic material residing inside the cell nucleus, that carry the genes from one generation to the next. As discussed below, single genes do not determine most behavioural adaptations, which instead can be modified by the actions of multiple genes and the environment. The genes form only a small fraction of the chromosomes that also carry a large amount of genetic material of unknown function. A single copy of all the genetic material from one organism is called a genome. Genomes are made up of DNA, which is quite simply the basic molecule of heredity. It consists of subunits we will be referring to as nucleotides. All genetic information in an organism is formed by multiple combinations of only four different nucleotides, designated according to the base that they carry: A (for adenine), G (for guanine), T (for thymine) and C (for cytosine). The products of a gene are RNA and proteins. To produce a protein, the first step is that of transcription. This is the copying of a DNA fragment to messenger RNA. RNA is ribonucleic acid, and it carries the code from the nuclear DNA to the second step, which is that of translation of the messenger RNA to synthesize amino acids. Amino acids are the building blocks for proteins that have multiple functions in an organism. Genes carry the genetic information, or code, for RNA and protein products that influence the performance or appearance of individuals. Therefore, behavioural or any other characters can only be transmitted genetically from one generation to the next through different combinations of genes and other fragments of genetic information, called regulatory regions, which control the amounts of genes produced. A phenomenon of interest to us is the occurrence of mutations. Mutations are changes in the DNA sequence, and they can provide for new genetic material to be transmitted. Mutations occur at random in any part of the genome and create different forms of genetic information at a certain position. Each alternate form of a mutation segregating in the population is called an allele. The impact of a mutation on a population is through their contribution to fitness for those animals that carry the mutations. As such, mutations can be favourable, neutral or unfavourable. By neutral, we mean that the mutation has no impact on any character, also called phenotype, or the impact it has on phenotypes has no association with the fitness of the animal with the mutation.

Many or Single Genes May Influence Behaviour

It has been shown that modifications in a single gene may produce large changes in behaviour. For example, as early as 1915, A.H. Sturtevant realized that a single mutation in the fruit fly *Drosophila melanogaster* that caused a change in eye colour could also affect mating behaviour (Brush, 2002). Another example is a single mutation in the monoamino oxidase (MAO) gene in humans that could cause an aberrant behaviour in males resulting in sexual attacks on the females from the same family, a behaviour called arsonism (Brunner *et al.*, 1993). The fact that a mutation in a single gene can cause a large modification in behaviour does not mean that it is the only gene involved. In other words, the MAO gene should not be called 'the aggression gene' simply because a mutation can have a large effect on this trait.

As an analogy, an automobile requires thousands of parts for the normal function of the engine. If any of the parts break down, for example the oil pump, the effect on the engine is dramatic and it will stop working, but we would not call the pump 'the engine of the car'. In the same way, single genes can drastically affect behaviour that is normally influenced by multiple genes. In general, a defined proportion of genes and environmental factors determine each behavioural trait. Figure 5.1A presents a trait in which 100% of the phenotype is determined by a single gene. These kinds of characters are called single gene traits, and they follow a simple (Mendelian) inheritance pattern. There is a complete correlation between the genetic information carried by the individual and phenotype expressed. In most cases, behavioural traits would instead belong to the category complex, or multifactorial traits, in which multiple genes in combination with environmental factors modify the phenotype. Figures 5.1B, C and D represent different proportions of genetic and environmental influence. A trait similar to the B example would be easier to dissect genetically, while the D example, with multiple genes contributing to a small proportion of the phenotypic variance, represent the most difficult situation for finding the underlying genes, and the phenotype for a certain genetic component is almost impossible to predict.

The difficulty in predicting the phenotype that one individual will have when multiple genes in combination with the environment interact can be illustrated by one example in which several, but not that many, genes are involved. Coat colour in mammals holds a particular fascination for all of us. Dogs have a wide variety of coat colours that are controlled by many alleles at several loci and the colour of the descendants of animals with certain genetic combinations are not always easy to predict. Lessons learned from the study of coat colour will prepare us for understanding the challenges of selection for more complex traits such as several behavioural traits. The genetics of coat colour in mammals is largely associated with several genes that modify pigment granules. This association can be either by altering their number, their shape, their arrangement or position in the hairs, or by substituting one type of pigment for another. The different coat colour loci

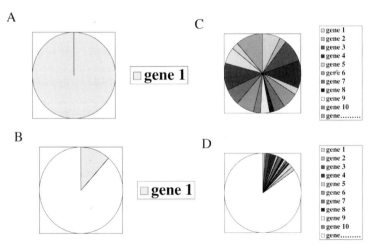

Fig. 5.1. A graphic representation of how a single trait may be affected by different numbers of genes. A represents a single-gene situation, where one gene completely determines a trait, whereas B–D show situations where different numbers of genes and environmental factors (the white area of the circles) contribute to various amounts of the phenotypic variation.

include: agouti, black and brown, albino, dilution and extension loci, greying and silver, merle, harlequin, tweed, spotting and ticking loci. These loci are either named by their impact on the phenotype relative to the colour we see, such as black or brown, their influence on the intensity we might see, such as the dilution, or in their distribution of colour patterns as we might see in merle, spotting or ticking. To make the matter more difficult, there are also a number of modifier genes that alter the phenotype that is typically expressed at other loci. There are breed differences for which loci are segregating and which are fixed, and there are also differences within breeds in the genetic background in which these genes are expressed. We must keep in mind that just the sheer number of these loci and possible genotypes within each of those loci makes studying coat colour a formidable task. It is therefore of extreme importance to select well-defined models to study the genetic contribution to behaviour and use powerful genetic tools to analyse these traits.

Behavioural Genetics in Dogs

Dogs present a quite unique situation for behavioural genetic analysis because they have been bred for centuries as much for their behaviour as for their looks. It is well known that dog breeds differ dramatically in appearance and size. However, the artificial selection imposed by humans was at least equally strong to develop breeds with particular capacities, such as herding, retrieval, pointing, guarding, resulting in groups with differences in activity, emotionality and aggressiveness. The contribution of genes to dog behaviours was clearly demonstrated by elegant breeding experiments performed over two decades by Scott and Fuller (Scott and Fuller, 1965). They studied five pure breeds – wire-haired fox terriers, cocker spaniels, basenjis, Shetland sheepdogs and beagles – and they also characterized all possible hybrids between two breeds. They studied a long list of canine habits, including for example investigative patterns such as walking with nose to ground, shelter building, grooming behaviour, dominance and submission, feeding, howling in unison or alone, chasing, barking, aggressive and fearful behaviour in response to human handling. It was quite clear that the five breeds presented large differences in several behaviours, particularly dominance between female–male pairs, tail wagging, lip licking, avoidance of human handler and biting. They also showed that these behavioural traits presented intermediate values for the hybrids between the five breeds, indicating a clear genetic contribution to the phenotypes.

A puzzle to solve is that the large behavioural differences between canid species contrasts with the similarities in the genome sequences of these two species. In fact, the genome sequence of dogs and wolves is almost identical (Vilá *et al.*, 1997; Kirkness *et al.*, 2003). The high sequence similarity between dogs and wolves suggests that altered gene expression, rather than changes in structural gene products, may be the main mechanism leading to trait differences between the two species. The idea that changes in regulatory regions have been a major motor for evolution was first proposed in 1971 (Britten and Davidson, 1971).

Subsequently, it has been postulated that mutations in regulatory sequences account for most biological differences between species (King and Wilson, 1975). Recently, these ideas have been supported through the search for the genetic basis of the difference in mental capabilities between humans and other primates, some of which share 98% sequence similarity (Chen and Li, 2001). In this case, altered gene expression, particularly in the brain, seems to be the main determinant for behavioural changes (Enard *et al.*, 2002; Gu and Gu, 2003). Pronounced changes in brain gene expression may not solely be associated with the development of human characteristics, they may also be important during the evolution of other species, as shown recently for dogs (Saetre *et al.*, 2004).

Altered gene expression can operate faster than other known evolutionary mechanisms leading to species differentiation, and alterations of gene expression patterns during the domestication of dogs may have triggered a rapid differentiation despite overall genome similarity. Furthermore, since selection for behavioural traits probably started earlier than selection for morphological traits (by the preferential breeding of docile and tame animals), the expression differences between dogs and wolves should be particularly pronounced in the brain, especially in tissues involved in emotion and behaviour. Different breeds of dogs should be more related genetically than dogs to wolves, which are generally accepted as two different species. However, morphological and behavioural differences among the breeds are extremely large (Hart and Miller, 1985). Studies on the genetic changes among different breeds have shown that genetic relatedness correlates with morphological similarity and geographical origins in some cases, while other genetic groups seem to correlate with behavioural traits (Parker *et al.*, 2004). Interestingly, canine breeds share fragments of the genome that segregate together from one generation to the next (called linkage disequilibrium blocks or LD blocks) that are 100 times more extensive than the LD blocks found among humans (Sutter *et al.*, 2004). Again here, breed-specific alterations in gene expression in the brain may be responsible for the large behavioural differences observed among breeds and the genetic differences responsible for expression changes may be located in LD blocks that differ between the breeds.

Different groups of dogs may not only be of importance for the study of normal behavioural differences among the breeds but also to dissect abnormal behaviours. It has been suggested that canine conditions of separation anxiety, obsessive-compulsive disorder, cognitive dysfunction, dominance aggression and panic disorder can be used as models for human generalized anxiety disorder, obsessive-compulsive disorder, Alzheimer's disease, impulse control disorders and panic disorder (Overall, 2000).

How Do We Know That a Behaviour Has a Genetic Component?

The first question that needs to be addressed before starting a large effort to find genes related to a behavioural trait is whether heredity has any importance.

There are several approaches and genetic experiments that can be used to estimate the extent to which observed differences among individuals are due to genetic differences and/or environmental differences. In general these methods are called quantitative genetics, and they estimate the extent to which differences among individuals are due to genes and environment, without specifying the number and type of genes and environmental factors involved. It is possible to estimate a parameter called heritability that indicates how much of the variability in a character that we observe is due to genes or environmental factors in general, without any specification of the number and type of genes or environmental factors involved. More technically, heritability estimates the amount of phenotypic variation in a trait (for example, behaviour) that is attributable to genetic variation in a specific population. In the case where no genetic component is present, the heritability is zero. In the other extreme case, the heritability is one when a single gene solely determines the character without any modification due to environmental factors. When heritability values are high, genetic differences are easier to identify using molecular genetic methods as explained below. However, the more genes involved, the less the contribution will be for each gene to the combined heritability value, and these genes will be more difficult to find.

Some of the methods used to study the involvement of genes are based on natural variation of behavioural traits. The crosses between different breeds of dogs to produce hybrids, as performed by Scott and Fuller as described above, are an example of this sort of strategy. Hybridization refers to the mating between representatives of different gene pools. The resulting progeny from the hybridization process is referred to as a hybrid. Hybrids are usually denoted as the F1 generation, the first filial generation. The main idea in these experiments is that if genes are involved in a certain trait, the crosses of individuals very different for that trait should produce hybrids with intermediate trait values (Fig. 5.2). Each original parental population (P1 and P2) has some variability in the trait, but all the hybrids produced (F1) have intermediate values. To reinforce the results, it is possible to do backcrosses: crosses between the F1 individuals and each of the parental populations (B1 and B2). If the trait has a genetic component, then the values of the backcross populations should be intermediate between F1 and P1 or P2, as shown in the figure. The example presented is a simple situation in which all the genetic components act in the same way, so that the presence of more than one component results in the sum of the effect of each separate component (additive effect). Other interactions are possible and different models can be built to understand the contribution of different genes.

Another type of method is artificial selection experiments, which provide the clearest evidence that genes are involved in behaviour. For example, foxes that were the tamest when fed and handled by humans were selected during multiple generations by a research group in Russia. The result of this artificial selection experiment is a new breed of foxes that are similar to dogs in their friendliness and eagerness for human contact (Trut, 2001). If a trait has a genetic component, the values of the population should be modified after selection in a

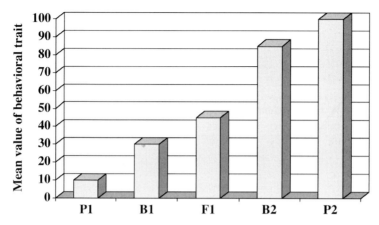

Fig. 5.2. A hypothetical case which demonstrates how crosses can prove a genetic control of behaviour. Assume that a behaviour differs between two parental populations (P1 and P2). If the behaviour is under genetic control, the offspring (F1) will have a trait value which is in between that of the parents. If F1 offspring are back-crossed to either P1 (producing B1) or to P2 (producing B2), the trait values of B1 will be intermediate to F1 and P1, and that of B2 intermediate to F1 and P2.

constant environment. If the trait is solely determined by environmental elements, the selection regime should not modify the average values for the trait in the selected population. One important observation is that selection for behavioural traits produces partial changes in one direction (for example, partial increases in average values) that continue steadily during multiple generations. This outcome strongly suggests that many genes contribute to variation in behaviour. If only one or two genes were involved, then the trait would change after a couple of generations and the values would stabilize without additional changes.

A problem of the two types of experiments described above is that they involve crosses of the studied populations. This may not always be practical and/or ethical. Instead, it is possible to study the correlation between the results of different behavioural traits measured in multiple individuals from different breeds of dogs. If a correlation is found, this would indicate common biological mechanisms and possibly common genes behind the selected traits (Saetre et al., 2006). Kenth Svartberg presents a similar approach in Chapter 11 of this book.

Identifying Genes

Once a genetic contribution to a trait has been established, it is possible to attempt the identification of the particular genes involved. As mentioned above, multiple genes and environmental factors determine most behavioural characters. Each gene, or rather genomic region that contains a gene included in a multiple gene system that can produce measurable changes in a trait, is called a quantitative trait locus (QTL). Unlike single-gene effects that are necessary and sufficient for the development of certain traits, each QTL has relatively small effects on the trait. Traditionally, genetic mapping has been used to localize QTLs in the genome. The classical analysis involves crosses between two strains with contrasting behaviours to produce hybrids (F1) and then inter-crosses between the hybrids to produce a second generation (F2). In this generation, recombination has produced genetically unique individuals. Recombination is the natural crossing over between the two copies of each chromosome that occurs during meiosis. This is important for mapping, because it breaks down the association between a QTL and molecular markers on the same chromosome. The result is that markers that are closer to a QTL tend to show greater association with the trait of interest than markers that are further away. At each position in the genome there are only two possible alleles (1 and 2) segregating from the parental strains. To map the position of a trait, one compares the values for the trait for each of the possible genetic combinations or genotypes (11, 12 or 22) for each marker. If there is a QTL close to the marker, the difference among the three genotypes will exceed chance expectations, that is, the QTL will be associated with a certain allele combination.

In an F2 cross there is only one generation in which recombination can break down the linkage between the QTL and nearby markers. As a result, mapping resolution is severely limited by sample size. The regions identified are usually very large, about 10–20 million base pairs of DNA, and such regions may contain thousands of genes among which the important behavioural gene may be located. Another problem is that multiple individuals usually have to be included in the analysis to increase the resolution and that replication of the results for genes of very weak effects is extremely difficult. Because of these difficulties, few QTL analyses have resulted in the identification of the gene contained in the genomic segment. One elegant exception is the identification of *RGS2* as a gene involved in anxiety in mice (Yalcin *et al.*, 2004). In this case, the authors used outbreed populations and they constructed combinations of alleles, called haplotypes, located close in the genome. Then they studied the probability that an ancestral haplotype carried a gene for anxiety. Finally, they used a test called quantitative complementation to evaluate the effect *RGS2*, a gene contained in the candidate haplotype, had on anxiety. Similar strategies could be used for the study of canine behavioural genetics. Recent studies of the genetic architecture of dog breeds have shown that individuals within one breed share quite large (1–3 million base pairs of DNA) regions of the genome (Sutter and Ostrander, 2004). As mentioned above, these regions are said to be in 'linkage disequilibrium' (LD) and they are called LD blocks. As LD in dog breeds is 20–100 times more extensive than in humans, a much smaller number of markers will be required for whole-genome association mapping studies in dogs compared to humans.

Transcriptome Analysis

Instead of looking for genomic regions that may contain QTLs for behavioural traits, an alternative approach for the identification of genes involved in behaviour is the analysis of segments of DNA that code for genes and are transcribed into RNA. The advantage of this approach is that current advances in molecular methods have made it possible to analyse simultaneously almost all the genes expressed in the genome using microarrays. The combination of all the genes from an organism that are expressed in a tissue at a certain developmental phase is called a transcriptome and the method is called transcriptome analysis. The main idea behind this analysis is that QTLs that code for behavioural genes should be transcribed, and differences in the amount of transcripts will in many cases be associated with small quantitative differences in gene function and behaviour. The notion that mRNA expression differences may have extreme importance in the function of genes is not new. As mentioned, changes in regulatory regions have been proposed as a major motor for evolution (Britten and Davidson, 1971). Subsequently, it has been postulated that mutations in regulatory sequences account for most biological differences between species (King and Wilson, 1975). It follows that many behavioural differences may also be caused by differences in expression levels of multiple genes.

To search for expression differences that may have an impact on behaviour, the brain is the central organ to analyse. Humans and primates, some of which share 98% sequence similarity, show multiple differences in expression levels in the brain, suggesting that altered brain gene expression may be the main determinant for behavioural changes that makes us 'human' (Enard *et al.*, 2002; Gu and Gu, 2003). Pronounced changes in brain gene expression may not solely be associated with the development of human characteristics, they may also be important for behavioural differences in other species, particularly those that were subject to rapid behavioural differentiation. For example, alterations of gene expression patterns during the domestication of dogs from wild wolves may have triggered a rapid differentiation despite overall genome similarity (Saetre *et al.*, 2004).

Brain expression differences may not only be responsible for behavioural differences between species but also between dog breeds with extreme differences in behavioural traits, and the investigation of affected genes may help the selection of candidates for additional functional characterization (Fig. 5.3). One caveat with this type of analysis is that genetic differences between the transcriptomes could be at least partially obscured by the confounding effect of expression differences produced by changes in the living environment of the species or breeds compared. For example, the diets of humans and apes are very different, and dogs are fed regularly with variable food types and live mostly in controlled environments, while wolves have a diet restricted to the availability of prey in the wild, and are subjected to long periods of fasting. This could have strong consequences in the hormonal balance and metabolism, with the possibility of strong consequences in the transcriptome balance of each species.

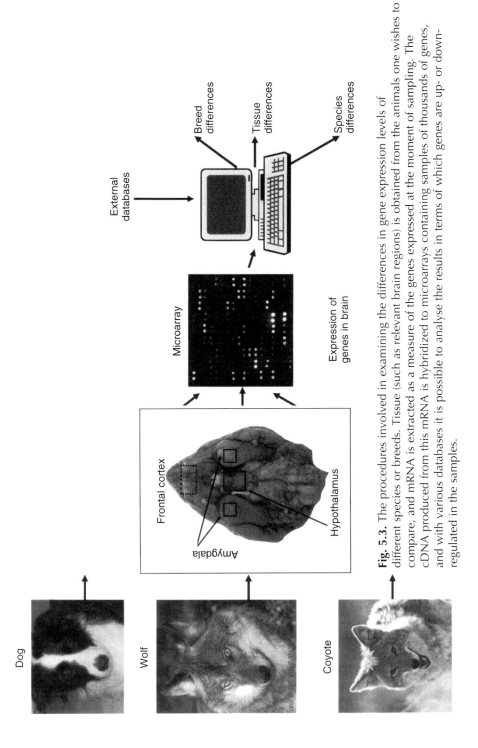

Fig. 5.3. The procedures involved in examining the differences in gene expression levels of different species or breeds. Tissue (such as relevant brain regions) is obtained from the animals one wishes to compare, and mRNA is extracted as a measure of the genes expressed at the moment of sampling. The cDNA produced from this mRNA is hybridized to microarrays containing samples of thousands of genes, and with various databases it is possible to analyse the results in terms of which genes are up- or down-regulated in the samples.

To overcome this difficulty, it is possible to study another species as a domestication model produced in a controlled environment. In the Russian farm-fox experiment, silver foxes (a colour morph of the red fox, *Vulpes vulpes*) were selected for non-aggressive behaviour towards man for more than 40 generations (Trut *et al.*, 2004). This resulted in silver foxes that not only were docile and friendly but also showed equally good skills as dogs in communicating with people (Trut *et al.*, 2004; Hare *et al.*, 2005). The transcriptome analysis of docile versus regular foxes indicated that many genes are modified due to environmental causes and only a few gene expression differences were the result of genetic differences between the groups (Lindberg *et al.*, 2005). Only genetic differences in gene expression are important for behavioural genetics and careful designs should be used to distinguish genetic from environmental changes in gene expression.

Mutations in the DNA that cause changes in gene expression are said to operate in *cis* or in *trans* mode. In other words, regulatory elements located close to the gene may be modified (*cis*-mutations) or other proteins that modify expression may be changed (*trans*-mutations). In general, the genetic basis of inter-individual variation in gene expression is simpler for *cis*-regulatory mutations (Cowles *et al.*, 2002; Rockman and Wray, 2002) and there is a growing body of evidence that that natural genetic variation can cause significant inter-individual variation in gene expression (Brem *et al.*, 2002; Rockman and Wray, 2002; Schadt *et al.*, 2003; Morley *et al.*, 2004; Chesler *et al.*, 2005). However, multiple regions of the genome may be involved in the regulation of the expression of a single gene and a network of complex interactions could be involved in the quantitative changes associated with behavioural traits.

Genetical Genomics: Potential for Behavioural Genetics

Genetical genomics is a neologism that refers to the combination of transcriptome analysis with QTL analysis aimed at dissecting the genetic regulation of gene expression (Jansen and Nap, 2001). In this case, QTL analysis as described above is performed using gene expression differences as traits. If the difference in expression is associated with a region of the genome located very close to the gene coding for expression, then a *cis*-mutation is probably involved. Conversely, if the QTL that affects expression is far away or in another chromosome, then a *trans*-mutation is involved. It should be noted that the use of microarray analysis enables the study of thousands of expression traits simultaneously. Some QTLs that affect expression levels will probably affect multiple genes (for example, QTLs coding for common transcription factors), while others will be more selective. This type of strategy has recently been used to analyse transcriptional networks in yeast (Hoeschele and Bing, 2005). If the expression profiling is carried out in the brain of animals for which behavioural tests have been conducted, there is a potential to identify QTLs that simultaneously affect behavioural response and gene expression.

A recent study of recombinant inbred (RI) mice from a cross of C57BL/6J and DBA/2J mice demonstrated the power of this approach. Using microarrays, Chesler and colleagues profiled genome-wide brain gene expression in 31 RI lines that had previously been genotyped and behaviourally phenotyped for ethanol-related behavioural traits (Chesler *et al.*, 2005). They were able to show that the expression level of the D2 dopamine receptor (Drd2) and several ethanol-related behavioural traits mapped to the same region on chromosome 9, which was located close to the position of the Drd2 gene itself. Although the Drd2 transcript is not polymorphic, several SNPs have been identified in its promoter, which suggests that a regulatory polymorphism in Drd2 is an important component of variation in ethanol-related behaviour.

A drawback of the genetical genomics approach is that few labs have the resources necessary to carry out the large-scale expression profiling of hundreds of F2 progeny needed for a genetic mapping project and only a few groups have combined genome-wide expression profiles with a genome scan of genetic markers in experimental populations (Brem *et al.*, 2002; Schadt *et al.*, 2003; Chesler *et al.*, 2005).

Functional Characterization of Selected Candidate Genes

All selected candidate genes need to be validated for their involvement in behavioural traits to close the circle. It will therefore be important to select appropriate behavioural tests for validation. The effect of environmental differences during testing has been intensely debated (Crabbe *et al.*, 1999). These results indicate that it may be impossible to reproduce exactly the same behavioural phenotypes in different laboratories. However, it should be possible to determine with certainty that a particular gene can cause qualitative changes in specific behaviours. Such qualitative changes may be extremely important for the differentiation of very closely related individuals such as different species that belong to the Canidae family.

The definition of a species being the most inclusive reproductive population is a useful one because it puts boundaries on gene flow. This means that members within a species can mate and produce viable offspring. For most cases, this definition is probably appropriate. The question is whether this definition is appropriate relative to the various species that make up the genus *Canis*. In fact, wolves, coyotes, jackals and dogs are inter-fertile, and crossbreeding still occurs in the wild. A better understanding of the genetics behind behavioural differences may help to distinguish man's best friends from their wild ancestors.

References

Brem, R.B., Yvert, G., Clinton, R. and Kruglyak, L. (2002) Genetic dissection of transcriptional regulation in budding yeast. *Science* 296, 752–755.

Britten, R.J. and Davidson, E.H. (1971) Repetitive and non-repetitive DNA sequences and a speculation on the origins of evolutionary novelty. *Quarterly Review of Biology* 46, 111–138.

Brunner, H.G., Nelen, M., Breakefield, X.O., Ropers, H.H. and van Oost, B.A. (1993) Abnormal behavior associated with a point mutation in the structural gene for monoamine oxidase A. *Science* 262, 578–580.

Brush, S.G. (2002) How theories became knowledge: Morgan's chromosome theory of heredity in America and Britain. *Journal of the History of Biology* 35, 471–535.

Chen, F.C. and Li, W.H. (2001) Genomic divergences between humans and other hominoids and the effective population size of the common ancestor of humans and chimpanzees. *American Journal of Human Genetics* 68, 444–456.

Chesler, E.J., Lu, L., Shou, S., Qu, Y., Gu, J., Wang, J., Hsu, H.C., Mountz, J.D., Baldwin, N.E., Langston, M.A., et al. (2005) Complex trait analysis of gene expression uncovers polygenic and pleiotropic networks that modulate nervous system function. *Nature Genetics* 37, 233–242.

Cowles, C.R., Hirschhorn, J.N., Altshuler, D. and Lander, E.S. (2002) Detection of regulatory variation in mouse genes. *Nature Genetics* 32, 432–437.

Crabbe, J.C., Wahlsten, D. and Dudek, B.C. (1999) Genetics of mouse behavior: interactions with laboratory environment. *Science* 284, 1670–1672.

Enard, W., Khaitovich, P., Klose, J., Zollner, S., Heissig, F., Giavalisco, P., Nieselt-Struwe, K., Muchmore, E., Varki, A., Ravid, R., et al. (2002) Intra- and interspecific variation in primate gene expression patterns. *Science* 296, 340–343.

Gu, J. and Gu, X. (2003) Induced gene expression in human brain after the split from chimpanzee. *Trends in Genetics* 19, 63–65.

Hare, B., Plyusnina, I., Ignacio, N., Schepina, O., Stepika, A., Wrangham, R. and Trut, L. (2005) Social cognitive evolution in captive foxes is a correlated by-product of experimental domestication. *Current Biology* 15, 226–230.

Hart, B.L. and Miller, M.F. (1985) Behavioral profiles of dog breeds. *Journal of the American Veterinary Medical Association* 186, 1175–1180.

Hoeschele, I. and Bing, N. (2005) Genetical genomics analysis of a yeast segregant population for transcription network inference. *Genetics* 170, 533–542.

Jansen, R.C. and Nap, J.P. (2001) Genetical genomics: the added value from segregation. *Trends in Genetics* 17, 388–391.

King, M.C. and Wilson, A.C. (1975) Evolution at two levels in humans and chimpanzees. *Science* 188, 107–116.

Kirkness, E.F., Bafna, V., Halpern, A.L., Levy, S., Remington, K., Rusch, D.B., Delcher, A.L., Pop, M., Wang, W., Fraser, C.M. and Venter, J.C. (2003) The dog genome: survey sequencing and comparative analysis. *Science* 301, 1898–1903.

Lindberg, J., Bjornerfeldt, S., Saetre, P., Svartberg, K., Seehuus, B., Bakken, M., Vila, C. and Jazin, E. (2005) Selection for tameness has changed brain gene expression in silver foxes. *Current Biology* 15, R915–916.

Morley, M., Molony, C.M., Weber, T.M., Devlin, J.L., Ewens, K.G., Spielman, R.S. and Cheung, V.G. (2004) Genetic analysis of genome-wide variation in human gene expression. *Nature* 430, 743–747.

Overall, K.L. (2000) Natural animal models of human psychiatric conditions: assessment of mechanism and validity. *Progress in Neuropsychopharmacology and Biological Psychiatry* 24, 727–776.

Parker, H.G., Kim, L.V., Sutter, N.B., Carlson, S., Lorentzen, T.D., Malek, T.B., Johnson, G.S., DeFrance, H.B., Ostrander, E.A. and Kruglyak, L. (2004) Genetic structure of the purebred domestic dog. *Science* 304, 1160–1164.

Rockman, M.V. and Wray, G.A. (2002) Abundant raw material for *cis*-regulatory evolution in humans. *Molecular Biology and Evolution* 19, 1991–2004.

Saetre, P., Lindberg, J., Leonard, J.A., Olsson, K., Pettersson, U., Ellegren, H., Bergstrom, T.F., Vila, C. and Jazin, E. (2004) From wild wolf to domestic dog: gene expression changes in the brain. *Brain Research. Molecular Brain Research* 126, 198–206.

Saetre, P., Strandberg, E., Sundgren, P.E., Pettersson, U., Jazin, E. and Bergström, T.F. (2006) The genetic contribution to canine personality. *Genes, Brain and Behavior* 5, 240–248.

Schadt, E.E., Monks, S.A., Drake, T.A., Lusis, A.J., Che, N., Colinayo, V., Ruff, T.G., Milligan, S.B., Lamb, J.R., Cavet, G., *et al.* (2003) Genetics of gene expression surveyed in maize, mouse and man. *Nature* 422, 297–302.

Scott, J.P. and Fuller, J.L. (1965) *Genetics and the Social Behavior of the Dog.* University of Chicago Press, Chicago, Illinois.

Sutter, N.B. and Ostrander, E.A. (2004) Dog star rising: the canine genetic system. *Nature Reviews. Genetics* 5, 900–910.

Sutter, N.B., Eberle, M.A., Parker, H.G., Pullar, B.J., Kirkness, E.F., Kruglyak, L. and Ostrander, E.A. (2004) Extensive and breed-specific linkage disequilibrium in *Canis familiaris*. *Genome Research* 14, 2388–2396.

Trut, L.N. (2001) Experimental studies of early canid domestication. In: Ruvinsky, A. and Sampson, J. (eds) *The Genetics of the Dog.* CAB International, Wallingford, UK, pp. 15–41.

Trut, L.N., Pliusnina, I.Z. and Os'kina, I.N. (2004) An experiment on fox domestication and debatable issues of evolution of the dog. *Genetika* 40, 794–807.

Vilá, C., Savolainen, P., Maldonado, J.E., Amorim, I.R., Rice, J.E., Honeycutt, R.L., Crandall, K.A., Lundeberg, J. and Wayne, R.K. (1997) Multiple and ancient origins of the domestic dog. *Science* 276, 1687–1689.

Yalcin, B., Willis-Owen, S.A., Fullerton, J., Meesaq, A., Deacon, R.M., Rawlins, J.N., Copley, R.R., Morris, A.P., Flint, J. and Mott, R. (2004) Genetic dissection of a behavioral quantitative trait locus shows that Rgs2 modulates anxiety in mice. *Nature Genetics* 36, 1197–1202.

Sensory Physiology and Dog Behaviour

Hermann Bubna-Littitz

Introduction

The purpose of this chapter is not to cover sensory physiology completely but to stress some topics of special behavioural relevance and/or which deviate markedly from the acuity of our own senses. The function of the sensory system is to provide information to the animal about its chemical, physical biotic and abiotic environment and of the state of the body of the organism itself. With this information, the animal is able to behave in an appropriate way to the changes in these factors, or to change the factors themselves. The effect of the behaviour is further monitored by the organism. If the behaviour was successful it will be stopped and if not another behavioural strategy will be tried. The uptake of information by the sensory system is not a passive process; on the contrary, it is an active one for two reasons:

1. The physical or chemical influences are transformed into generator potentials and/or action potentials. These neurophysiologic parameters allow the processing of the input.
2. The sensory system is influenced by the state of the organism. For example, a dog which is hungry will react with predatory behaviour on seeing a rabbit with higher probability than a satiated dog. This means that motivation exerts an influence on the sensory system by filtering the stimuli and altering the attention for special types of stimuli (Bubna-Littitz, 2005).

The sensory system plays a crucial role not only in the lives of each individual, but also in intra- and interspecific communication. Dogs use acoustic, chemical (odours, pheromones) and visual signals. A signal is defined as behaviour assigned with a distinctive meaning, which is recognized by the receiver (Tembrock, 1987), and this obviously relies heavily on a sensory system which is able to detect the signal.

Development of the Sensory System and of Behaviour

The development of the sensory system is more or less finished at the end of the third week of the dog's life (see also Chapter 4 in this volume). From the 28th day of life onwards a transition from variable locomotor responses to adult-like postural and equilibratory abilities occurs. Visual and auditory recognition of littermates and other individuals is developed. In addition, approach and avoidance behaviours emerge (Fox, 1965). In accordance with the development of the sensory and motor system, the volume of the brain increases rapidly until the end of the 4th week of life. With the end of week 7, the gain of volume ceases (Fox, 1965). The function of the sensory system is the precondition for imprinting (see also Chapter 8 in this volume). Imprinting is a learning process that is only possible within a short period of development, called the 'sensitive period'. This process assigns a distinct meaning to an object of the biotic or abiotic environment, e.g. the object of sexual or social behaviour. The effect of imprinting is irreversible.

Hearing

Basic anatomy and physiology of the auditory organs

The auditory system consists of the external, middle and inner ear (see Fig. 6.1). The external ear (earlap; pinna) is a structure consisting mainly of cartilage. This provides the breed-specific form of the external ear and the muscles which can rotate the pinna, thus enabling better detection of the direction from which the sound is coming. The pinna works not unlike an ear trumpet. It tunnels the sound waves (pressure waves of the air) to the middle ear. The border between the external and the middle ear is the tympanic membrane, which is thin and elastic, with a size of about 25 mm^2. The sound waves cause an oscillation of this membrane, which is transduced by three small bones (incus, malleus and stapes). The last of these three ossicles is in tight connection with the oval window, which is the entrance to the inner ear.

Since the area of the tympanic membrane is about 20 times bigger than the area of the oval window, the force exerted here is 20-fold higher than the one exerted on the tympanic membrane. In the middle ear, two of the smallest muscles of the body are situated: one of them is connected with the stapes and the other one with the tympanic membrane. The contraction of these muscles decreases the transfer of vibration through the middle ear. A reflex elicits the contraction when the incoming energy of the sound is too high. This reflex is also

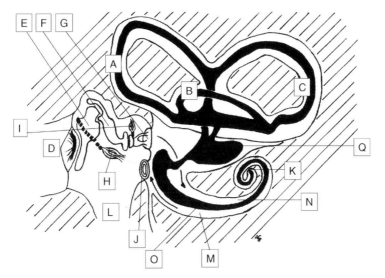

Fig. 6.1. The inner ear. A, B, C: semicircular canals; D: tympanic membrane; E: malleus; F: incus; G: stapes; H: musculus tensor tympani; I: musculus stapedius; J: foramen rotundum; K: top of the cochlea; L: auditory (Eustachian) tube; M: lower tympanic tunnel; N: upper tympanic tunnel; O: scala media; Q: utricle.

activated immediately before a dog starts to bark and therefore offers protection to its own ear.

From the oval window, a pressure wave is running up the upper tunnel of the cochlea, which contains fluid (perilymphe). At the top of the cochlea, there is a connection to the lower tunnel, so that the pressure wave finally arrives at the round window, closed by an elastic membrane, which bulges out when the wave arrives. That is the precondition for the propagation of the wave, since fluids are not as compressible as air.

Between the upper and lower tunnel there is the middle tunnel (scala media) which contains the Organ of Corti. This consists mainly of sensory cells, which react on vibration. High frequencies are detected in the beginning of the scala media, and low frequencies at the end (at the top of the cochlea).

The neural action potentials arising from the cochlea arrive in the cochlear nuclei in the medulla oblongata in the brain. From there, these are transmitted by various brainstem routings to different neuronal structures (e.g. the cerebellum and the auditory cortex), partly crossing over from one side to the other. The conscious perception of sound occurs in the auditory cortex. The connection of the auditory system with the cerebellum is responsible for eliciting the so-called shoulder-reaction, whereby the individual looks over the shoulder towards the origin of the sound. The location of the source of sound may be detected in two ways. First, by comparing the intensity of the sound coming from the right and left ear in the auditory cortex, where information from both sides is available. Second, the shift in phase between the sound waves arriving in the left and right ear can be detected in the superior olivary nuclei ('place mechanism'; Pearson and Pearson, 1976).

Hearing in dogs

Dogs have an excellent auditory system. They can locate the source of sound far better than man: dogs can distinguish two different sources of sound when the angle between the connecting lines (head-source) is as small as 1° 26′. In man the corresponding value is 4° 18′ (Buddenbrock, 1952). This ability is very important in predatory behaviour performed in complex terrains (e.g. wood with scrub). The direction can be detected by motions of the earlap and by using the difference of phase of the sound waves arriving at the left and right ear.

The dog is able to hear ultrasound (frequencies higher than 20 kHz) up to 60 kHz (Overall, 1997). What is the benefit of this ability? In accordance with Peters and Wozencraft (1989), one can assume that puppies communicate in distress situations with the mother by means of ultrasound. We know from physics that the damping of an oscillation correlates negatively with its frequency. Therefore, ultrasound is used when the signal should not be heard at greater distances to avoid possible predators detecting it. In predatory behaviour, it could play a role in finding prey, since some rodents use ultrasound in communication too.

Fox (1965) distinguishes between the following types of canine vocalizations:

1. Infantile sounds like crying, whimpering and whining.
2. Howling. The interpretations of this type of sound are controversial. In wolves, two functions are discussed: to get in contact with other members of the pack and to mark the territory by sound. In some dogs, howling is elicited when the owner is absent.
3. Aggressive growling.
4. Submissive whining.
5. Territorial defensive barking.

From the acoustical signal, the receiver dog can presumably estimate the size of the vocalizing dog. Dogs range in body mass from chihuahua to Saint Bernard, a 100-fold difference. In addition, the vocal tract length differs tremendously and, because of that, also the acoustic signal (Riede and Fitch, 1999).

Vision

Basic anatomy and physiology of the visual organs

The eye of the dog shows an anatomy similar to that of humans. The white layer encasing most of the eyeball is called the sclera (see Fig. 6.2). A part of the anterior is modified to be transparent, a part called the cornea. When light goes through the cornea, it enters the anterior chamber, penetrates the lens and arrives at the

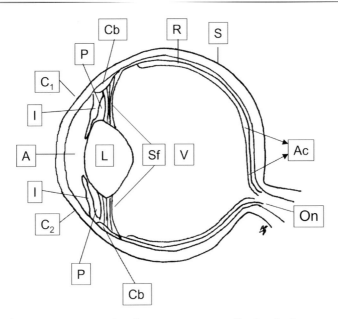

Fig. 6.2. The eye. A: anterior chamber; Ac: area centralis; C_1, C_2: between C_1 and C_2 is the cornea; Cb: ciliary body; I: iris; L: lens; On: optic nerve; P: posterior chamber; R: retina; S: sclera; Sf: suspensory ligaments; V: vitreous humour.

posterior chamber. Both chambers are filled with fluid. The iris forms the border between these two chambers. By contraction or dilatation of smooth muscle fibres, the diameter of the pupil ('the central hole of the iris') can be changed and thus the amount of light reaching the retina. The lens is elastic. It is fixed by suspensory ligaments, which originate from the circular ciliary muscle. The contraction of this muscle decreases the tension of the suspensory ligaments. Because of the inherent elasticity of the lens, it becomes more spherical and has more light refraction that focuses the image of nearer objects on to the retina. The retina is in the rear part of the eye, and mainly consists of a layer of photoreceptors and neurones processing the information in a first step towards visual awareness. Behind the retina lies a light reflecting pigment called tapetum lucidum. Light, which has passed the photoreceptors, is reflected, and passes the photoreceptors again. This layer is especially important in nocturnal animals and is responsible for the 'night shine' of the eyes in dogs and cats, for example. In the dog's retina, there are two types of photoreceptors: one type, the rods, can only differentiate the intensity and not the wavelength (colour) and have high sensitivity. Therefore, this type is responsible for vision when light intensity is low, like in the evening or night. The second type is the 'cones', which have lower sensitivity but can discriminate different wavelengths. Primates, including humans, can distinguish three basic colours (they are trichromatic): one cone type is sensitive for red, one for green and one for blue. Dogs only have two subtypes, so they are bichromatic.

Vision in dogs

Unlike humans, the dog has no fovea centralis but an area centralis. As in the fovea centralis, that is where the highest density of photoreceptors is found. The effect of this is that in this area the resolution of optical signals is higher than in the rest of the retina.

The importance of vision depends to some extent on the breed of the dog: it is less important in breeds such as bassets, beagles and bloodhounds, which track prey using their noses ('scent hounds'). It is of more importance in dogs hunting in free areas like Afghans and greyhounds, which have good sight ('sight hounds'). However, in general, visual acuity is higher in humans than in dogs.

The colour vision of the dog has been a matter of discussion for years. As already mentioned, dogs (like many other mammals) have only two types of cones. Some therefore believe that dogs are unable to distinguish different colours in the same way as humans do. However, in an experiment using a discrimination task, it was clearly shown that dogs can differentiate colours. The dogs had to identify the designated correct colour out of magenta, cyan and yellow. Finding the correct colour was rewarded by food, and the dogs could clearly distinguish all three colours (Antolini-Messina, 1996).

In another experiment also using a discrimination task, it was found that the ability to distinguish different scales of grey is poorer than that of humans (Pretterer *et al.*, 2004), but that the dogs could readily discriminate between red, green, blue and two types of turquoise and all scales of grey. None of these animals was able to distinguish between a third type of turquoise (wavelength: 480 nm) from the corresponding grey shade. Therefore 480 nm is the so-called neutral-point for the dog (a colour with this wavelength looks like grey) (Pretterer, 2000).

The behavioural aspect of vision in the dog is that it plays a major role in finding prey. Predatory behaviour presumably makes use of a cascade of stepwise sensory information. By olfaction, the dog finds the direction where possible prey could be found. With the help of vision and hearing it finds the prey and the instinctive killing behaviour is then released. Visual signals also play an important role in intra- and interspecific communication: the posture of the whole body, the mimic of the face, ears, mouth, muzzle, the fur (flat or raised), the tail held upwards or downwards or wagging (see Fig. 6.3a–c). An alert dog is depicted in Fig. 6.3A: the ears are erect, there is a gentle slope in the hip, a relaxed tail and no piloerection can be seen. When a dog's fear level increases, the head is lowered, the ears are pulled back and become limper. The lips become looser and the eyebrows are arched, the tail is lowered or between the hind limbs (Fig. 6.3B). When aggression increases the ears are raised, the fur piloerected, the teeth shown, nostrils and pupils are widened, the tail is upright or wagging, and the hind limbs stretched out (Fig. 6.3C). The submissive dog, on the other hand, makes itself look smaller and in extreme situations, it rolls on its back, flexes its limbs and feet, exposes its belly and tucks in its tail.

In some breeds with great deviations from the appearance of the wolf, problems in sending visual signals can occur. For example, dogs with hanging ears or

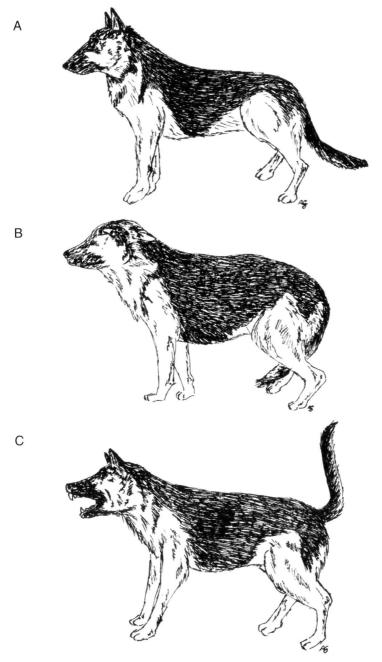

Fig. 6.3. Signal behaviour of the dog. The figure shows three dogs in different states of mind: an alert one (A), a fearful one (B) and an aggressive one (C). From the pictures one can derive that the posture differs tremendously, thus another dog can recognize the mood of a conspecific from a far distance.

short snout may have problems sending appropriate signals and sometimes may be misunderstood by other dogs. Another example is the chow chow, which normally has stretched hind limbs, which is misinterpreted by some dogs as demonstrative behaviour.

The Olfactory System

Basic anatomy and physiology of the olfactory organs

The olfactory sensory cells lie in the brain in the olfactory bulb. These have two axons (nerve fibres), and are therefore called 'bipolar sensory cells'. One axon goes to the mucous membrane of the nose and sprouts there. These sprouts have about 1000 different receptors for various types of chemical compounds (Buck, 2000). The other axon has contact with other neurons in the olfactory bulb and a first stage of the processing of olfactory information happens there. Thus, the sensory neurons of the olfactory system are the only ones that have direct contact with the brain. The sensation of scent is the result of neuronal processing: from the sensory neurons, information is transferred to the olfactory bulb situated on the other side of the brain, and to the limbic system, which influences the emotional state of the organism. Axons from the olfactory bulb conduct information to the olfactory cortex as well. The typical scent of another conspecific or something else is therefore due to the neural processing of receptor information of many chemical compounds with different structure and concentration.

Olfaction in dogs

In free-ranging dogs or wolves, the olfactory system is important to detect prey. The dog can usually detect the direction where the scent is coming from. In doing so, the moisture of the nose may be important, since it may allow the dog to identify the wind direction.

The olfactory system of the dog is extremely sensitive compared to that of humans. This fact is not only the consequence of the larger area of the olfactory epithelium of the dog compared with humans (5 cm^2 in humans and 150 cm^2 in the German shepherd) (Brüggemann *et al.*, 1965), but also of the higher density and sensitivity of the sensory cells. For some compounds, the sensitivity of the dog's olfactory system is millions of times higher than in men (Brüggemann *et al.*, 1965). Hence, it is clear that smells play a much more important role to the dog than to humans: when a male dog meets a female, olfactory cues provide information about her state in the sexual cycle, for example. Each individual dog can probably be identified by its personal olfactory profile. Therefore, a change of the individual odour of a dog can cause problems, since it might not be identified by another dog: change of smell can turn a dog into an 'alien' individual. Drugs used for anaesthesia can change the personal odour, a problem well known in cats.

Dogs, mainly males, mark their territories with urine. It contains information about sex and perhaps about state in the ranking order: generally the alpha-male raises the hind limb extremely high, so that the urine is deposited higher than for lower ranking males. The posture during urination demonstrates the position in the ranking order. In neutered female dogs the rising of the hind limb during miction is sometimes observed too.

Dogs also use defecation for marking. Some individuals try to deposit their faeces as high as possible, for example on fences or trunks. The scratching of the ground after defecation is rather an imposing behaviour than a dispersion of faeces or odours. This behaviour is more extensive when another male passes. In this situation, sometimes the scratching dog starts to growl.

Dogs used for hunting or used in the military or police service are trained to find deer, persons, etc., by sniffing. Here, it is necessary to distinguish between primary and secondary tracks: the primary track is the scent of the person; plants and insects for example, which have been destroyed by the footsteps of the searched person or prey, cause the secondary track and deliver additional scents. Thesen *et al.* (1993) distinguish three phases during tracking the footprints of a person. In the first phase (searching phase), the dog is going straight forward while sniffing close to the ground. When it has found the track, it identifies the direction the person was walking by comparing the intensity of the scent of the first footprints (phase 2, deciding phase). During tracking (tracking phase, phase 3), the pattern of sniffing is similar to that of phase 1. When tracking, the dog sniffs with a frequency of six inhalations per second for about 15 times and then a respiration follows. In phase 2, the respiration follows each period of 35–50 sniffs.

In dogs searching for drugs the decision phase (phase 2), which is the most strenuous one, might be the only one occurring. After about 20 minutes of work, the dog is usually exhausted. Sniffing leads to an increase of the energy needed for inspiration and expiration, which can be demonstrated by blood samples (Strasser *et al.*, 1993). One can speculate that pack hounds overcome the problem of exhaustion by alternating with each other in tracking. As a speculation, I want to put forward another hypothesis: scent hounds, like bassets, beagles and bloodhounds, track their prey using their noses. The breeds mentioned have very long, hanging ears. Maybe these serve a function in whirling up the scent and guiding it to the nostrils.

Pheromones

Pheromones are chemical compounds that are produced from one animal and influence the behaviour of a conspecific. These play a large role in sexual behaviour. For example during proestrus and oestrus, a pheromone (methyl-*p*-hydroxybenzoate) is produced in the vagina of the dog, and this compound stimulates the interest of the male in the female. The pheromones are not only detected by the olfactory system described above but also by the vomeronasal organ or accessory olfactory system. This organ is best described as a fluid-filled tubular structure which opens into the nasal cavity via a duct at its anterior end (Buck, 2000). The axons from the sensory neurones of that organ conduct action potentials to the accessory olfactory bulb and from there these potentials are transmitted, amongst others, to the hypothalamus but not to the olfactory cortex – this means that the dog does not 'feel the smell' of a pheromone. The hypothalamus is a structure involved in the regulation of the secretion of hormones, some of them involved in sexual behaviour and reproduction. We should remember that the olfactory system is in tight connection with the limbic system, which influences the emotional state of the organism.

Taste

Basic anatomy and physiology of the taste organs

Taste is the least understood of the human senses (Rouhi, 2001). For a long time it has been a white landmark in physiology. Compounds that can be tasted are detected by taste cells clustered in taste buds. These buds can be found on the tongue, palate, epiglottis and the upper third of the oesophagus (Buck, 2000). However, the sensation of taste is a complex of many different sensory inputs, as has long been recognized.

> The great majority of the sensations we call taste, however, are in reality complex sensations, into which smell and even touch largely enter. When the sense of smell is interfered with, as when the nose is held tightly pinched, it is very difficult to distinguish the taste of various objects. An onion, for instance, the eyes being shut, may then easily be confounded with an apple.
>
> (Huxley, 1881)

Even in recent textbooks of physiology you will find the statement that there exist only four types of taste receptors, being the source of the perception of sweet, sour, bitter and salty. In recent years, a fifth receptor type has been detected: the umami receptor (Kurihara and Kashiwayanagi, 2000).

> Compounds with sodium or hydrogen ions, respectively, are perceived as salty and sour. Carbohydrates are generally associated with sweetness, although other compounds may elicit the same sweet perception. The alkaloids caffeine and quinine are quintessential bitter, but likewise many other compound types taste bitter. Umami is the savoury taste frequently associated with protein-rich food such as meat and cheese.
>
> (Rouhi, 2001)

Very probably, new types of receptors will be found and some may not be present in all individuals. Some humans have receptors for PROP (6-n-propyl-2-thiouracil) and have the perception of extreme bitter taste. For persons without that receptor, the PROP does not elicit any impression at all (Rouhi, 2001).

The information from the taste receptors is transmitted to the olfactory cortex, where the information processing takes place. The intensity of the different inputs (tastes) is compared and therefore the perception of a distinct taste like 'pepper' occurs. That inputs of the olfactory system are integrated is reasonable – remember the citation of Huxley mentioned above. The information from the taste receptors is not only transferred to the cortex via the thalamus, but also to the nucleus of the solitary tract. From the solitary nucleus arises, amongst others, the vagal nerve. This fact could explain the phenomenon that in the presence of distinct tastes vomiting occurs, since the vagal projects directly to parasympathetic and sympathetic neurons in the medulla and spinal cord that mediate various autonomic reflexes (Saper, 2000).

Taste in dogs

In the dog, taste functions similarly to in humans (Kurihara and Kashiwayanagi, 2000). In dogs, there are five types of receptors too. For the dog, the receptor sensitive to umami is probably very important, for dogs eat meals with high protein content. However, as already stated by Huxley, cited above, taste is a function of olfaction. Taking into account the tremendous sensitivity of the dog's olfactory system, we can assume that it has a much greater range of taste impressions than a human does. Olfaction seems to be more important in dogs than taste: if a dog smells a preferred substance, this can easily be used to trick it to eat another, non-preferred substance (Overall, 1997). The main purpose of olfaction and taste presumably is to select the correct food and to avoid harmful or poisoning ones.

Pain

Basic anatomy and physiology of the pain sensory organs

> The sensations we call pain – prickling, burning, aching, stinging and soreness – are the most distinctive of all the sensory modalities. Pain is of course, a submodality of somatic sensation like touch, pressure, and position sense, and serves an important protective function: it warns against injury that should be avoided or treated.
>
> (Basbaum and Jessell, 2000)

Pain can be elicited by different qualities of stimuli, e.g. heat, coldness, pressure and deterioration of tissue. During the first sensation of pain ('the sharp pain'), it can easily be localized. Nerve fibres with high conduction velocity (A-delta fibres) transmit this sharp pain. The second sensation – sometimes a burning one – is due to action potentials transmitted by nerve fibres with low velocity of conduction (C fibres). This sensation cannot be localized as well as the sharp pain. Seen from the point of evolution, the nonmyelinated C fibres occurred before the myelinated A-fibres. The explanation could be that exact localization of the source of pain only makes sense when the organism can react exactly, which needs a certain state of evolutionary development. Free nerve endings serve as nociceptors (receptors for painful stimuli) and conduct the information by C fibres. These are activated by chemical, thermal and physical energy and therefore are called polymodal nociceptors. There are specialized nociceptors, just like mechanical and thermal receptors. These use A delta fibres for transmission.

The action potentials from the nociceptors and the sensory neurones are transmitted to the dorsal horn of the spinal cord. There, a first step of information processing occurs: when someone feels pain, for example in an arm, and scratches himself, pain is reduced. During scratching, mechanoreceptors are activated and transmit action potentials by A beta fibres to the spinal cord. These potentials decrease the conduction of the potentials stemming from the C and A delta fibres (gate control theory; Basbaum and Jessell, 2000). This could explain why a dog sometimes bites its tail or paws when in pain.

The nociceptive information is transferred to the brain along five ascending pathways (spinothalamic tract, spinoreticular tract, spinomesencephalic tract, cervicothalamic and spinohypothalamic tract). Seen from the behavioural point of view, the most interesting one is the spinomesencephalic tract: it transmits information to the thalamus, a brain structure which among its functions has elicitation of aggression. The thalamus is also part of the limbic system, which plays a role in influencing mood and emotion, and it conducts information to the mesencephalic reticular formation too, which is responsible for the arousal reaction. This view of functional neuroanatomy explains what we might observe when a dog is hurt during sleep: immediately on waking up it may attack the person or animal nearby, which in the dog's opinion may be the reason for the pain. If the pain is lasting for extended time, the mood is severely reduced. The last processing of nociceptive information takes place in the cerebral cortex and leads to conscious perception of pain.

The threshold of pain increases during stress. States of stress may be caused by negative painful stimuli, but similar physiological states may be caused by positive stimuli too (see also Chapter 4 in this volume). This makes some functional sense: under the condition of stress, a normal reaction to pain would be a risk to the animal's life: stopping during flight because of hurting a paw could result in death. Therefore, when stress occurs, hormones are secreted which increase the pain threshold. These hormones belong to the group called 'endorphins' (endogenous morphine). This fact may explain why dogs fighting with each other or attacking a person can only be stopped by extreme physical force.

Pain in dogs

In dogs, thresholds of pain are very variable. The height of the threshold depends partly on the breed of the dog (e.g. working breeds of dogs vs toy breeds; The American College of Veterinary Anesthesiologists, 2003).

One widely discussed question is whether pain is inflicted in puppies during tail docking, i.e. amputation of a part or all of an animal's tail. Commonly, surgical methods are used. Dogs usually are docked within an age of 3–5 days without anaesthesia. In some breeds, docking is required by the breed standards. In an excellent review, Bennett and Perini (2003) state: 'Pups do exhibit those pain responses of which they are capable, and there is every reason to expect that they experience considerable pain while being docked'. These authors argue that a puppy can feel pain since the second sensation of pain (burning pain, see above) is due to C fibres, which are unmyelinated. Myelination of the other nerve fibres is not finished at the age of 3–5 days. Taking into account the above described gate control theory, one might expect that the puppy feels more pain than an adult dog, since the pain-reducing effect for example of the a-Beta fibres is not working, because these are not fully myelinated in puppies of that age.

Acknowledgements

I want to thank Univ. Prof. Irene Sommerfeld-Stur and Dr Hellmuth Wachtel for the stimulating discussions on this topic and Andreas Grätzl for the excellent illustrations.

References

Antolini-Messina, M. (1996) Untersuchungen über das Farbsehvermögen des Hundes. DVM thesis, The Veterinary University of Vienna, Austria.
Basbaum, A.I. and Jessell, Th.M. (2000) The perception of pain. In: Kandel, E.R, Schwartz, J.H. and Jessell, Th.M. (eds) *Principles of Neuronal Science*, 4th edn. McGraw-Hill, New York, pp. 472–491.
Bennett, P.C. and Perini, E. (2003) Tail docking: a review of the issues. *Australian Veterinary Journal* 81, 208–218.
Brüggemann, J., Horn, V., Moustgaard, J., Hill, H., Kment, A. and Spörri, H. (1965) *Scheunert/Trautmann, Lehrbuch der Veterinärphysiologie*, 4th edn. Paul Parey, Berlin.
Bubna-Littitz, H. (2005) Verhaltensphysiologie. In: Engelhardt, v.W. and Breves, G. (eds) *Physiologie der Haustiere*, 2nd edn. Enke Verlag, Stuttgart, Germany, pp. 644–649.
Buck, L.B. (2000) Smell and taste: the chemical senses. In: Kandel, E.R., Schwartz, J.H. and Jessell, Th.M. (eds) *Principles of Neuronal Science*, 4th edn. McGraw-Hill, New York, pp. 625–647.
Buddenbrock, W.v. (1952) *Vergleichende Physiologie*, Vol 1 Sinnesphysiologie. Birkhäuser Verlag, Basel, Switzerland.
Fox, M.W. (1965) *Canine Behavior*, 1st edn. Charles C. Thomas, Springfield, Illinois.
Huxley, Th.H. (1881) *Lessons in Elementary Physiology*, 6th edn. Macmillan and Co., London.
Kurihara, K. and Kashiwayanagi, M. (2000) Physiological studies on umami taste. *Journal of Nutrition* 130 (4 Suppl.), 931S–934S.
Overall, K.L. (1997) *Clinical Behavioral Medicine for Small Animals*, 1st edn. Mosby, St Louis, Missouri.
Pearson, R. and Pearson, L. (1976) *The Vertebrate Brain*, 1st edn. Academic Press, London.

Peters, G. and Wozencraft, W.C. (1989) Acoustic communication by fissiped carnivores. In: Gittleman, J.L. (ed.) *Carnivore Behavior, Biology, and Evolution*, 1st edn. Cornell University Press, Ithaca, New York, pp. 14–56.

Pretterer, G. (2000) Untersuchungen über das Helligkeits- und Farbsehvermögen des Hundes. DVM thesis, The Veterinary University of Vienna, Austria.

Pretterer, G., Bubna-Littitz, H., Windischbauer, G., Gabler, C. and Griebel, U. (2004) Brightness discrimination in the dog. *Journal of Vision* 4, 241–249.

Riede, T. and Fitch, T. (1999) Vocal tract length and acoustics of vocalization in the domestic dog (*Canis familiaris*). *The Journal of Experimental Biology* 202, 2857–2867.

Rouhi, M.A. (2001) Unlocking the secrets of taste. *Chemical and Engineering News* 79, 42–46.

Saper, C.B. (2000) Arousal, emotion, and behavioural homeostasis. In: Kandel, E.R., Schwartz, J.H. and Jessell, Th.M. (eds) *Principles of Neuronal Science*, 4th edn. McGraw-Hill, New York, pp. 873–888.

Strasser, A., Hochleithner, M. and Bubna-Littitz, H. (1993) Zur Belastung von Gebrauchshunden bei der Suchtgiftsuche. *Wiener Tierärztliche Monatsschrift* 352–355.

Tembrock, G. (1987) *Verhaltensbiologie*, 1st edn. Gustav Fischer Verlag, Jena.

The American College of Veterinary Anesthesiologists (2003) American College of Veterinary Anesthesiologists' position paper on the treatment of pain in animals. http://www.acva.org/professional/Position/pain.htm

Thesen, A., Steen, J.B. and Doving, K.B. (1993) Behavior of dogs during olfactory tracking. *Journal of Experimental Biology* 180, 247–251.

7 Social Behaviour of Dogs and Related Canids

Dorit U. Feddersen-Petersen

Why Study Social Behaviour in Dogs?

Just like their ancestors, dogs are highly social animals. Some intriguing insights into their social life may be achieved by studying both dogs themselves and their close relatives. In my group, we have spent a long time studying more or less social canids (*C. aureus* L. (golden jackals), *C. latrans* Say (coyotes), *C. lupus* L. (wolves)) and domestic dogs (*C. lupus* forma *familiaris*) of various breeds tackling questions dealing with social development, social communication, social organization, play and aggression.

The animals lived under comparable semi-natural conditions, where group size, sex ratio and so on were kept constant. Closely related canids, such as domestic dogs of various breeds and their progenitor, the wolf, when kept under comparable living conditions, show conspicuous similarities, but also a number of differences in social behaviour and its ontogeny – with marked intraspecific variability. Hence, comparative studies of social canids and dogs offer excellent opportunities to record constant traits with regard to the development and significance of individual or species-typical behaviour particularities – as well as those induced by domestication and breeding (Feddersen-Petersen, 2004).

Pack-living wolves are social canids par excellence. They develop a very high degree of sociality, a fact that may be judged as a kind of pre-adaptation for domestication: many capacities fit with the very high degree of social contacts and interactions of humans, living in reproductive units (families).

Wolves show a variety of facial expressions and body postures, while for many dog breeds the possibility to communicate precisely is lost due to an extreme diversity in morphological characters. Regarding the mimic area, in many dog breeds we find only fragments of the wolf's diversity, fine details and gradations. In brachycephalic breeds, the forehead is always wrinkled, as is the nose area, and teeth baring often is not possible because of prominent flews. Thus, several facial regions and a lot of signals have been lost for communication. The channel of acoustic communication is partly hypertrophic in all breeds analysed so far. Categories of function/emotion expressed include social play, play soliciting, exploration, care-giving, social contact and 'greeting', fear and agonistic behaviour. Interactions range from mildly agonistic biting of infants by adult dogs to affiliative acts like grooming. Via bark differentiations, the dog vocalizations seem to have developed into an increasingly communicative component of social interactions. In our studies, the German shepherds and especially the bull terriers were found to bark extraordinarily often and in a variety of social contexts. Furthermore, sounds occurring in different phonetic qualities were more common in domestic dogs than in wolves. The evolution of the barking system could be a parallel to the vocalizing human social partner.

Studies of social behaviour in canids may not only be interesting from the point of view of communication biology. Tomasello and Call (1997) summarized their review of primate cognition by noting that 'The experimental foundation for claims that apes are "more intelligent" than monkeys is not a solid one'. While some authors (Flack and de Waal, 2000) focus on nonhuman primates as the most likely animals to show precursors to human morality, others have argued that we might learn as much or more about the evolution of human social behaviour by studying social carnivores (Bekoff, 1995).

According to Bekoff (2000), comparative data on social behaviour in canids may broaden the study of animal sociality. For example, when more or less social canids (wolves, domestic dogs, coyotes, golden jackals) engage in social play, they appear to expect to be treated fairly by their conspecifics. While golden jackals (*Canis aureus* L.) usually show play signals when rough play sequences are turning into aggressive encounters, in wolves (*Canis lupus* L.) play signals punctuate longer lasting social plays as a kind of 'play markers' (sensu Bekoff, 2000). Social canids like wolves also learn rules to coordinate their lives. Furthermore, pack size in wolves is regulated by social factors, which will be clear from examples given in the continuation of this chapter.

Domestic Animals with Special Demands

Dogs represent a 'special kind' among domestic animals, as many breeds or forms of them have been intimate social partners to humans for many thousands of years. The unique, intimate relationship between dogs and humans reaches back 15,000 years BP at least (Savolainen *et al.*, 2002; see Chapters 2 and 3 in this volume), some breeds even apparently preferring humans to conspecifics.

Dogs are the oldest domestic animals. In agreement with morphological, ethological and chromosomal data, recent genetic findings (Tsuda *et al.*, 1997; Vilà *et al.*, 1997; Randi *et al.*, 2000; Savolainen *et al.*, 2002; see Chapters 2 and 3 in this volume) indicated that wolves and dogs are closely related, wolves being the only ancestors of domestic dogs. Dogs are highly variable in appearance, and this is also true for behavioural traits. They are adapted to the ecological niche of 'Hausstand' (a German word describing life among humans) (Herre and Röhrs, 1990). Dogs are frequently treated in a highly anthropomorphic way on the one extreme hand or objectified to serve human's vanity on the other.

I want to discuss some comparative data on social behaviour in dogs and related canids to detect specific peculiarities of dogs, to promote a better understanding of their minds. Dogs cannot be characterized without humans. Therefore, in the margin, some socio-ecological explanations for the evolution of human sociality will be offered in this chapter. Social carnivores' social behaviour and organization resembles that of early hominids in a number of ways (as pointed out by Schaller, 1969), and analogue traits exist. Just like humans, wolves cooperate within packs, and dogs easily form working relationships with man. Humans and social canids therefore harmonize within a variety of social accounts. This appears to have been instrumental in the domestication process.

Dogs are very sensitive to human signals, for example, to a person's 'body language', while human abilities to interpret dog signals vary. Given the closeness and loyalty of the human–dog relationship in so many aspects, the social-cognitive abilities of dogs are largely focused on humans. Dogs have acquired a skill for communication 'with humans in a unique way' (Hare *et al.*, 2002; see Chapter 12 in this volume).

In our research, we performed extensive comparisons between European wolves (*Canis l. lupus* L.), coyotes (*Canis latrans* Say), golden jackals (*Canis aureus* L.) and domestic dogs (*Canis lupus* f. *familiaris*, various breeds). Behaviour studies were first carried out under environmental conditions that resemble those of the ancestral species: so-called 'semi-natural', variable housing conditions, where animals can largely live according to their own free choice. Regarding dogs, that means observing them in wide open-air enclosures while they are allowed to live in packs, comparable to the wolf pack or the social groups of coyotes and golden jackals. 'Normal behaviour' of domestic animals is usually defined as the behaviour of healthy animals living in this reference system of a 'semi-natural' environment.

The Social Behaviour of Canids – an Overview

Social development, social communication, social organization, play and aggression in coyotes and golden jackals differ significantly from other canids. Whereas wolves and domestic dogs may show what has been called an 'exaggerated' degree of sociality (Lewin, 1987), coyotes may best be characterized as 'social animals that exhibit solitary attitudes' (Macdonald, 2004). Coyotes are therefore highly adaptable. However, they do not possess the wolf's sociability. This is reflected in their more stereotyped expressive behaviours. This is even more obvious in golden jackals. Coyotes and jackals have fewer behavioural elements, and fewer possibilities of combining signals to differentiated postures or faces. The amount of social play in golden jackals, coyotes, wolves and domestic dogs (various breeds) varies.

Social behaviour in dogs and wolves

As mentioned earlier, the high sociality of pack-living wolves may constitute a kind of pre-adaptation for domestication. The main characteristic patterns of social behaviour in wolves (and dogs) are the following traits:

- Social play: during play, basic rules of social life are learnt: how hard they can bite, how roughly they can interact and how to resolve conflicts.
- Sense of fairness (following Bekoff, 2001; De Waal, 2003): while playing, codes of social conduct are learnt. No matter what the functions of play may be, there seems to be no doubt that it has benefits and that the absence of play can have devastating effects on social development (as remarked by Power, 2000). In Bekoff's opinion, a sense of fairness is common to social canids (and other animals) because there could be no social play without it, and without social play individual animals and entire groups would be at a disadvantage. In this sense fairness is adaptive, it helps animals to survive in their particular social environment, and individual animals benefit from this behaviour.
- Communicative skills: these are numerous and highly differentiated. There are a variety of displays in wolves and dogs which have to be learnt (by associating the performance with the consequences). Complexity within the dog's vocal repertoire is achieved by many subunits of bark, standing for specific motivations, information and expressions. Complexity within the dog's vocal repertoire is extended by the use of mixed sounds in the barking context. Transitions and gradations occur to a great extent via bark sounds: harmonic, intermediate and noisy subunits (as defined below).
- Family groups (packs): social systems vary from pairs and family groups to packs. Pack size is influenced by food supply and by social factors (Mech, 1970). In wolf packs there may be division of labour, and food sharing.
- Long-term monogamy: this is the most common mating system, but polyandrous mating occasionally occurs. Mating behaviour of dogs (especially pedigree dogs) is nearly exclusively controlled by man. Thus, for the most part there is no possibility for dogs to live in a long-term relationship with a sexual partner. Some observations support long-term monogamy in some dogs (feral dogs, dingoes).
- Bonding behaviour: in wolves and dogs, this can be much stronger and longer lasting than in species closely related to ourselves, such as chimpanzees.
- Parental care (care of other adults and siblings): provisioning of cubs is done by both parents and all other pack members act as helpers. Helpers are common in dingoes (Australian dingo and the so-called New Guinea singing dog) and occur in several dog breeds.

- Social learning from conspecifics: in wolves the duration of the period of exposure to parental influence (thus the opportunity for social learning) is long, and normally many pups remain in the pack permanently. There they have the possibility to learn social behaviour, hunting skills and feeding behaviour.
- Communicating social rights, dominance hierarchy: a differentiated communication is correlated with a complex social hierarchy. Dominance always characterizes specific relationships and does not refer to individual characteristics (Van Hooff and Wensing, 1987). When establishing and maintaining a stable social rank, wolves and dogs perform agonistic encounters in a unidirectional way. If dominant behaviours, for example a so-called T-sequence or a high posture, are alternately performed by both interacting animals, this demonstrates that the interacting individuals have not achieved a stable rank order. Dominance implies the acceptance of restrictions by the subdominant animal. Dominant wolves strikingly often initiate dominant behaviours, such as dominant body postures or mimics, when they engage with pups. In the case shown in Fig. 7.1c–e, muzzle-biting (with pronounced biting inhibition) follows

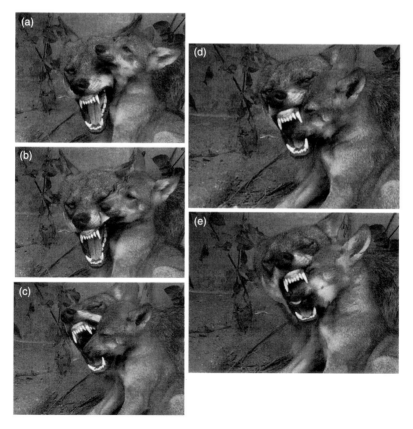

Fig. 7.1. Communicating social rights to infant wolf.

immediately on licking the bridge of the elder wolf by the pup (b) and placing the head on the adult wolf's snout (a). The pup performs a kind of T-sequence 'from snout to snout' (a), thus, testing a 'high social status' in a certain behavioural context.

A missing or unstable rank order can be indicated by individuals exhibiting ambivalent communications when interacting. Such a behaviour is controlled by two different motivations (e.g. attack and fear) occurring simultaneously or in short, rapid succession. It may produce expressions including 'incongruous' signals: in Fig. 7.2 a juvenile wolf (left side) demonstrates partly dominance-indicating behaviours (muzzle biting, nose-wrinkled, pawing the alpha male on the head) in combination with submissive gestures/signs (folded ears laying back, flat forehead). Ambivalent expressions are common behavioural traits in wolves and not exclusive of a particular social status. In fact, higher levels of ambivalence were observed in high-ranking individuals, the alpha when interacting with any other male and the beta when interacting with the omega (Fig. 7.2). According to this finding, a dog's body posture during agonistic episodes should never be taken as the only element to establish a functional diagnosis of aggression (Fatjo *et al.*, 2006).

Some Details of Dog Sociality

Although dogs descended from wolves fairly recently in evolutionary terms, their social behaviour has changed extensively, particularly in some breeds. In our research, we found that some dog breeds are unable to cooperate (in a very basic

Fig. 7.2. Play-biting/mimic play in 6-week-old wolf pups.

manner: just doing things together) and compete in groups, reflected in difficulties in establishing and maintaining a rank order (e.g. poodles). The interactions in these dog groups are not functional, and the members have difficulties coping with challenges from the environment. It is striking that tactical variants of conflict solving (to appease, animate or inhibit the opponent), a common practice in wolves, do not exist in groups of several dog breeds. These strategies, however, are important for pack maintenance. Within many groups of dogs, trivial conflicts often escalate into damaging fights.

This may lead to acute stress becoming chronic. Other breeds (Nordic breeds: Alaskan malamutes, Siberian huskies, samoyeds and German shepherds, hunting dogs, some terriers, Fila Brasileiros) have better means to cope with social conflicts. These dogs are mentally in a good shape when living under 'semi-natural' conditions.

We found that breed groups differed mainly with respect to the following behavioural measures (Feddersen-Petersen, 2004):

1. There was much more frequent and much more severe aggressive behaviour in toy poodles, West Highland white terriers, Jack Russell terriers, some bull terriers and some labrador retrievers.
2. There was much more frequent and variable social play in the wolves, standard poodles, the earlier mentioned Nordic breeds, German shepherds, Filas and weimaraner hunting dog, which were able to show aggressive encounters in a more ritualized manner.
3. There was greater social tolerance, more nonagonistic approaches and more allogrooming in the wolves, malamutes, huskies, samoyeds, German shepherds, some bull terriers, American Staffordshire terriers, golden retrievers and e.g. Fila Brasileiros.

Coyotes and golden jackals showed very different behavioural strategies, and could readily cope with challenges from the environment. The possible reasons for the decreased capability of several dog breeds to form stable groups will be discussed later. However, we can already note here that poodles, for example, will interrupt every/any interaction with conspecifics, if/whenever a human appears, an intriguing fact regarding the social minds of these dogs. In German shepherd dogs, the duration of a threat display before and following an attack was constant – whereas in wolves pre-escalative and post-escalative threat display dyads differed significantly in duration. These findings suggest that during the process of domestication, dogs have lost essential capabilities necessary to adapt to living exclusively with conspecifics in groups. Nordic breeds (Alaskan malamutes, Siberian huskies, samoyeds) were exceptions, as mentioned above.

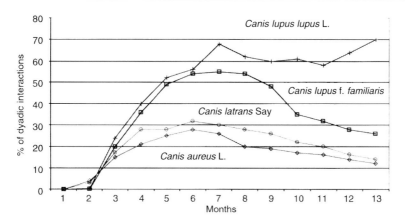

Fig. 7.3. Frequency of social play in some related canids during their first year of life.

Development of Social Behaviour in Wolves, Domestic Dogs and Related Canids

During the first year (and beyond), juvenile wolves are characterized by pronounced play activities (Fig. 7.3). In addition to the well-known canid forms of social play, e.g. contact and chase games, which normally mix one into one another and consist of highly variable sequences, wolves display a further category: communication or facial expression games. Indeed, these games make up a large part of their social play (Feddersen-Petersen, 1991). Nineteen per cent of all social behaviour patterns recorded in the third month of life involve mutual communication of playfully expressed, exaggerated signals almost exclusively from the facial area (Fig. 7.4).

Contact games, or bite and fight games (or play-fighting) are the earliest type of play. Transitions to serious conflicts rarely happen in wolves. After the third month, social play is carried out with pronounced role assumption and subsequent reversal: a puppy will first play, for example, the 'underdog' by sending a few, limited body signals indicating social inferiority in a playful, exaggerated manner, while its partner plays the role of the social superior. The 'social superior' may then switch roles and bring elements of submission (sensu Schenkel, 1967) into the play. Playful exaggeration refers to the amplitude and speed of expressive movements or behaviour patterns. Contact games normally appear in combination with chase games in which the primary goal is the playful pursuit of one's opponent. During play, inequalities in size, strength and dominance between playmates are reduced, and the cooperation and reciprocity that are essential for a longer lasting play occur. The role of social play for cooperation is more than a hypothesis. The rights and wrongs of social interaction can be learnt best in a play atmosphere. Biting inhibition and partner preferences (preferring to

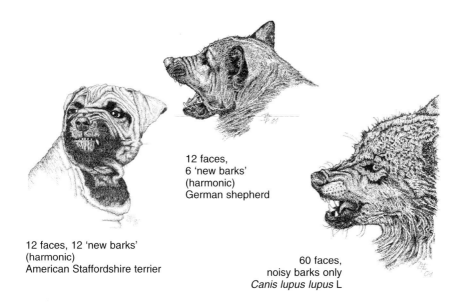

Fig. 7.4. Do reduced facial expressions correlate with sophisticated barkings?

initiate as well as to accept play signals) are important for forming social bonds between certain animals.

Solitary play is much less common as compared to social games. Domestic dogs exhibit a maximum in play activity between the sixth week of life and approximately 6 months of age (Fig. 7.3). At this time, a relatively stable hierarchy has been established and from now on the games are tinted by a primarily aggressive tone and may even end in serious fights (for example, in poodles). Play-fighting predominates, but decreases rather abruptly towards the eighth month and continually thereafter. In contrast to wolves, many dogs (living in groups) play a great deal less after the fifth or sixth month.

While wolves primarily integrate visual signals into the games, most dog breeds develop bark-games (Zimen, 1981), i.e. they initiate play by 'play-sounds' (harmonic barks, growling, vibrato-sounds). Wolves start to play earlier and play frequency is almost always greater than the corresponding 'dog-norm'.

Agonistic interactions appear earlier in most dog groups than among wolves. Agonistic behaviour among wolves living in an age-structured family group is determined structurally by a high degree of ritualization. Conflicts between wolves are mainly seen in the form of complex, ritualized fighting. Dogs, as a rule, fight in a much less ritualized manner. For example, an attack launched by a dominant poodle male escalated into grabbing and bite-shaking in 70% of the observed cases, regardless of the opponent's reaction. Group aggression was sometimes seen, in which many group members joined a collective attack on a threatened animal. The social distance between the high-ranking animal or animals and the rest of the group appears large among dogs.

In golden jackals pronounced aggression is present during the first 4–6 weeks of life in groups of pups. Bite and fight games, the sole form of social play in golden jackals, regularly escalate into unritualized conflicts, including serious biting (intended to harm). The frequency of agonistic behavioural patterns reaches its maximum at approximately 9 weeks. At this age the litters have established certain hierarchies and groupings or pairs have created spatially divided subgroups. Intragroup play frequency increases, while intergroup cohabitation in a relatively restricted area is regulated by expressing ritualized threat motions. The frequency of play increases within the subgroups (Fig. 7.3) and social games free of aggression take place up to the onset of the first oestrus period. The expression elements of these games are restricted almost entirely to the agonistic context.

Communication games are lacking in golden jackals. Wide mouth-opening is shown as an intention motion shortly before playful biting in the course of play-fighting (and only here) and is not answered as a play-signal through facial exaggeration. The relaxed, open-mouth face of golden jackals is always displayed as a prelude to play-fighting.

Solitary play is more common than social play up to 3 months of age. The total amount of play is considerably less than that of wolves at a comparable age.

The frequency of agonistic interactions in coyotes is considerably above the wolf and dog norms and is especially pronounced prior to the establishment of social groups (at approximately 9 weeks of age) and to the onset of the first oestrus. Play signals (like bows) often initiate plays.

In coyotes and jackals, stereotyped play signals (very clear and striking) predominate (approx. 70% of play interactions), while wolves and dogs have a variety of fine-graded play signals, play faces and play vocalizations (dogs, especially play barks), which serve as 'play markers', announcing longer lasting play sequences.

In all canids there is an increase in the frequency of play at 1–3 months, the socialization period, during which intraspecific (and interspecific) attachments are formed (Scott and Fuller, 1965).

Why Do Several Dog Breeds Have Problems in Forming Stable Groups?

In adult wolves we found about 60 different distinct facial expressions (Fig. 7.4). We fixed 11 'facial regions' (including 2–13 different signals), resulting in an enormous number of possible displays. In dogs, face morphology differs in the various breeds, which reduces the number of possible expressions. In addition, visual signals and body postures are less numerous, less differentiated and reduced in amplitude as well as in clearness, compared to the wolf's diversity. Thus, expressions are reduced. Reductions in facial expression produce problems in social communication, as we have found in mixed poodle–wolf groups, and may cause increased aggression. Finally, dogs are adapted to living circumstances with

Fig. 7.5. Fighting play in pugs.

***Canis lupus* L:**
A single, short sound of the noisy type, occurring solely within aggressive contextual categories: threat, attack, warning, defence and protest (Schassburger, 1993)

Canis lupus* f. *familiaris
Individual dogs have specific barks with a range of meanings (Feddersen-Petersen, 2000; Young, 2002). For example: a heterotypic sequence of harmonic barking (1), and harmonic (2,3) / intermediate (4) growl (vibrato) as a play soliciting signal

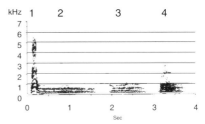

Fig. 7.6. Barking of *Canis lupus* L. and barking in domestic dogs.

humans. As we found out, aggressive encounters decrease if humans participate more frequently in group life.

Barking in Wolves and Dogs

Barking in wolves is rare, occurring most often in the social context of fighting. Wolves' bark sounds are of the noisy type (characterizing agonistic encounters) (Fig. 7.6).

Fig. 7.7. Ambivalent communication in juvenile wolf (left).

In dogs there is an abundance of barks/barkings in highly differing social contexts. Barking as a whole seems to be both directed to humans and an adaptation to living with humans.

The principal method, used in our studies of vocalizations, was sonographic analysis. This facilitates the identification of sounds and reveals whether subjective classifications can be verified by objectively measured parameters. Finally, meanings, functions and emotions were examined for all major sounds identified and were analysed in terms of relationships between sound structure and signal function, signal emission and social context as behavioural response, and overlapping channels of communication. This provides a model for future vocalization studies in domestic dogs, a model which draws parallels between structure, motivation, emotion and ontogeny to arrive at a more comprehensive understanding of sound systems in various dog breeds.

According to Schassburger (1993), the wolf's vocal repertoire consists of 11 basic sound types. In the Alaskan malamute and Tervueren breeds, the vocal repertoires turned out to be very similar, but meanings, function and emotions, behavioural responses and overlapping channels of communications varied much more. The vocal repertoire of the bull terrier turned out to be highly hypertrophic and much more differentiated in some close-range vocalizations, such as barking and growling (Fig. 7.6, sonogram on the right side). This applies for most of the breeds we have analysed so far (standard poodles, toy poodles, Kleine Münsterländer, weimaraner hunting dogs, Tervueren, American Staffordshire terriers, Jack Russell terriers, bull terriers, West Highland white terriers, German shepherds, Alaskan malamutes, Siberian huskies).

American Staffordshire terriers, which have markedly reduced facial expression, emitted 12 'new barks' (Fig. 7.4), i.e. barks which do not occur in wolves.

Harmonic barks were found in all dog breeds analysed so far. Alaskan malamutes emitted two harmonic barks in play situations, and when contacting another animal (or a human being) in a slightly submissive pose (with stimulative expression). Thus, complexity within the dog's vocal repertoire, and therefore enhancement of its communicative value, is achieved by many subunits of bark (and growl), representing specific motivations, informations, expressions and communications. Complexity within the dog's vocal repertoire is extended by the use of mixed sounds in the barking context. Transitions and gradations occur to a great extent via bark sounds: harmonic, intermediate and noisy subunits.

We have attempted to analyse dog barks in categories of function and emotion. These categories include social play, play soliciting, exploration, caregiving, social contact and 'greeting', loneliness (loneliness-bark) and agonistic behaviour (offensive/defensive threat). 'Interaction' was the most common category of social context for most vocalizations. Interactions ranged from mildly agonistic biting of infants by adult dogs to affiliative acts like grooming. In addition, there are acoustic signals following or indicating distress via bark sounds or gradations including barks.

Dogs and Humans

Social canids may be characterized by an 'exaggerated degree of sociality' (Lewin, 1987). During domestication, dogs seem to have adapted their social behaviour to a large extent to the life with humans, representing to a far extent their ecological niche. But the domestication of social cognition in dogs counteracts their social communication with conspecifics. Dogs changed in so many ways during domestication and now depend on humans for their benefit. In so many respects, modern dogs seem to be better adapted to communication with humans than with other dogs.

References

Bekoff, M. (1995) Play signals as punctuation: the structure of social play in canids. *Behaviour* 132, 419–429.
Bekoff, M. (2000) Social play behaviour. Cooperation, fairness, trust and the evolution of morality. *Journal of Consciousness Studies* 8, 81–90.
Bekoff, M. (2002) *Minding Animals: Awareness, Emotions, and Heart.* Oxford University Press, New York & London.
De Waal, F.B.M. (2003) Reconciliation among primates: a review of empirical evidence and unsolved issues. In: Mason W.A. and Mendoza, S.P. (eds) *Primate Social Relationships.* State University of New York Press, New York, pp. 111–114.
Fatjo, J.F., Feddersen-Petersen, D.U., Ruiz De La Torre, J.L., Amat, M., Mets, M., Braus, B. and Mantacea, X. (2006) Aggression in wolves: ambivalent behaviour as a model for comparable behaviour in dogs. *Applied Animal Behaviour Science.*

Feddersen-Petersen, D. (1991) The ontogeny of social play and agonistic behaviour in selected canid species. *Bonn, Zoologische Beiträge* 42, 97–114.

Feddersen-Petersen, D. (2000) Vocalization of European wolves (*Canis lupus lupus* L) and various dog breeds (*Canis lupus* f. *fam.*). *Archives of Animal Breeding, Dummerstorf* 43, 387–397.

Feddersen-Petersen, D.U. (2004) Communication in wolves and dogs. In: Bekoff, M. (ed.) *Encyclopedia of Animal Behavior*, Vol. 1. Greenwood Press, Westport, Connecticut, pp. 385–394.

Flack, J.C. and de Waal, F.B.M. (2000) Any animal whatever: Darwinian building blocks of morality in monkeys and apes. *Journal of Consciousness Studies* 7, 1–29.

Hare, B., Brown, M., Williamson, C. and Tomasello, M. (2002) The domestication of social cognition in dogs. *Science* 298, 1643–1636.

Herre, W. and Röhrs, M. (1990) *Haustiere – zoologische gesehen*. Gustav Fischer, Stuttgart, Germany.

Lewin, R. (1987) The origin of the modern human mind. *Science* 236, 668–669.

Macdonald, D.W. (2004) Die große Enzyklopädie der Säugetiere (ed.) Könemann in der Tandem Verlag GmbH, Königswinter.

Mech, L.D. (1970) *The Wolf: The Ecology and Behavior of an Endangered Species*. The Natural History Press, New York.

Mech, L.D. (1999) Alpha status, dominance, and division of labor in wolf packs. *Canadian Journal of Zoology* 77, 1196–1203.

Power, T.G. (2000) *Play and Exploration in Children and Animals*. Lawrence Erlbaum Associates, Hillsdale, New Jersey, pp. 48–49.

Randi, E., Lucchini, V., Christensen, M.F., Mucci, N., Funk, S.M., Dolf, G. and Leschke, V. (2000) Mitochondrial DNA variability in Italian and East European wolves: detecting the consequences of small population size and hybridization. *Conservation Biology* 14, 464–473.

Savolainen, P., Zhang, Y., Luo, J., Lundeberg, J. and Leitner, T. (2002) Genetic evidence for an East Asian origin of domestic dogs. *Science* 289, 1619.

Schaller, G.B. and Lowther, G.R. (1969) The relevance of carnivore behavior to the study of early hominids. *Southwestern Journal of Anthropology* 25, 307–314.

Schassburger, R.M. (1993) Vocal communication in the timber wolf, *Canis lupus*, Linnaeus. *Advances in Ethology* 30.

Schenkel, R. (1947) Ausdrucksstudien an Wölfen. *Behavior* 1, 81–129.

Schenkel, R. (1967) Submission: Ist features and functions in the wolf and dog. *American Zoologist* 7, 1–27.

Scott, J.P. and Fuller, J.L. (1965) *Genetics and the Social Behavior of Dogs*. University of Chicago Press, Chicago, Illinois.

Tomasello, M. and Call, J. (1997) *Primate Cognition*. Oxford University Press, New York.

Tsuda, K., Kikkawa, Y., Yonekawa, H. and Tanabe, Y. (1997) Extensive interbreeding occured among multiple matriarchal ancestors during the domestication of dogs: evidence from inter- and intraspecies polymorphisms in the D-loop region of the mitochondrial DNA between dogs and wolves. *Genes and Genetic Systems* 72, 229–238.

Van Hooff, J.A.R.A.M. and Wensing, J. (1987) Dominance and its behavioral measures in a captive wolf pack. In: Frank, H. (eds) *Man and Wolf: Advances, Issues and Problems in Captive Wolf Research*. Junk, Dordrecht, pp. 219–252.

Vilà, C., Savolainen, P., Maldonado, J.E., Amorim, I.R., Rice, J.E., Honeycutt, R.L., Crandall, K.A., Lundeberg, J. and Wayne, R.K. (1997) Multiple and ancient origins of the domestic dog. *Science* 276, 1687–1689.

Yin, S. (2002) A new perspective on barking in dogs (*Canis familiaris*). *Journal of Comparative Psychology* 116, 189–193.

Zimen, E. (1981) *The Wolf: A Species in Danger.* Delacorte Press, New York.

Zimen, E. (1982) A wolf pack sociogram. In: Harrington, F.H. and Paquet, P.C. (eds) *Wolves of the World: Perspectives of Behavior, Ecology and Conservation.* Noyes Publications, Park Ridge, New Jersey, pp. 282–32.

Learning in Dogs

Pamela Reid

Introduction

While the principles of learning have been derived primarily from studies of animals in laboratories, nowhere are they more thoroughly revealed than in our work with the domestic dog. Dogs have been selectively bred to serve many functions for humankind, most of which require some degree of trained behaviour (Coppinger and Coppinger, 2001). In recent years, we have relied less on dogs for their working ability and more for their companionship, yet dog training has become increasingly sophisticated as aficionados strive to create ever more complex and precise behavioural routines. The trend for dogs to function as household pets and family members is an additional force driving the infiltration of learning theory into conventional dog training and behaviour modification, because humans often find aspects of dog behaviour objectionable (see Chapter 13 in this volume). Illustrations of fundamental learning processes abound from these intentional efforts to control and manipulate dog behaviour.

Learning Processes in Dogs

A fascinating, and often frustrating, enterprise is to map correspondence of learning theory on to the complications of the real world. Highly specific aspects of learning can be delineated and explored within the controlled environment of the laboratory but this rarely happens in the 'messy' world of dog training and behaviour modification. Throughout this chapter, I provide examples of particular learning principles; yet bear in mind that in virtually all situations, multiple processes interact to produce learning.

Non-associative learning

Two of the most fundamental forms of learning are non-associative in nature. By that, I mean that the dog is not forming associations between events, rather the dog is learning about events that appear to occur without connection to other events. For instance, imagine staying at a hotel situated underneath the flight path for a nearby airport. As the first airplane thunders overhead, your dog pins back its ears, tucks its tail and dives under the bed. It takes a few moments after the noise ends for the dog to emerge, only for it to hustle back to its hiding spot at the first hint of the next plane. Remarkably though, as the day wears on, you notice that the dog no longer hides, in fact, it appears not even to hear the planes any longer. The dog has *habituated* to the sound. Habituation can occur in response to any repeated stimulus (Thompson and Spencer, 1966). In essence the dog learns, through repetitions, that there is no significance to the sound and, therefore, there is no need to attend to the stimulus. Dogs often habituate to everyday events – the dog that moves with its family into an apartment building initially barks at every noise in the hallway, but eventually learns to ignore all the irrelevant sounds. Habituation is very specific: introduce a novel similar stimulus and the dog will react all over again (Thompson and Spencer, 1966). Habituation is also somewhat temporary. If the airport switches to a distant runway for a few days before resuming use of the one by the hotel, the dog will startle again the first few times a plane flies over. Quickly, however, the magnitude of the dog's response will decline (Leaton, 1976).

Occasionally, a dog can be exposed to presentations of a stimulus, say fireworks, and, instead of habituating, the dog's response to the stimulus actually intensifies. With each ensuing exposure, the dog becomes more reactive. This phenomenon is termed *sensitization*. Dogs are more likely to sensitize when the stimulus is especially intense, such as fireworks, gunfire, thunder and other loud percussive sounds (Davis, 1974). Trainers who use electronic stimulation to provide incentive for learning find that dogs sometimes sensitize to the shock. A dog that startles mildly the first time it receives a shock might show a much more intense response, jumping, yelping, or in rare cases even biting the handler, after a few more presentations. Unlike habituation, sensitization is a more generalized phenomenon (Miller and Domjan, 1981); often a dog that has sensitized to shock will also visibly startle to other external events, like a light touch, a blowing leaf or a spoken word.

Operant conditioning

Operant or instrumental conditioning is a form of associative learning, meaning that associations or relations are established, specifically relations between behaviour and its outcomes (Skinner, 1938). Animals frequently engage in behaviour that produces changes in the environment: behaviour followed by a pleasant event is more likely to be emitted again; while behaviour followed by unpleasant events is less likely to be emitted again (Thorndike's Law of Effect) (Thorndike, 1911). Operant behaviour is often described as intentional, voluntary or goal directed; the animal chooses to perform the behaviour because in the past the behaviour was instrumental in producing certain consequences (Salzinger and Waller, 1962). Much of dog training consists of operant conditioning. A dog learns to attend to its name because the owner, after saying the dog's name, frequently takes the dog for a walk; a dog learns to sit when the owner speaks to it because the action of sitting sometimes results in a tasty biscuit; a dog learns to jump on the counter for the opportunity to snatch the butter dish; a dog learns to avoid jumping up on the counter because the action results in the owner shouting and shooing the dog off.

Behavioural outcomes can be classified into four categories, based on the behaviour–outcome contingency: (i) positive reinforcement, in which a behaviour produces an appetitive stimulus (a pleasant thing); (ii) positive punishment, in which a behaviour produces an aversive stimulus (an unpleasant thing); (iii) negative reinforcement, in which a behaviour causes the cessation or prevents the delivery of an aversive stimulus; and (iv) negative punishment (also known as omission training), in which a behaviour causes the cessation or prevents the delivery of an appetitive stimulus (Domjan, 2003). Positive reinforcement and positive punishment are both characterized by a positive contingency between the behaviour and the outcome (the behaviour occurs and some event happens), negative reinforcement and negative punishment are both characterized by a negative contingency between the behaviour and the outcome (the behaviour occurs and some event ends or is prevented). The two types of reinforcement result in the behaviour being more likely in the future; the two types of punishment result in the behaviour being less likely in the future.

Examples of each of these contingencies abound in dog training. The dog lies down. This action results in the owner tossing a favourite ball; if the behaviour of lying down increases, lying down has been positively reinforced. The dog chews the leg of the couch. This action results in the owner scolding 'no' and rapping the dog on the nose; if chewing the couch ceases, chewing the couch has been positively punished. The dog jumps up at the owner at the front door. This action results in the owner ignoring the dog when the dog was attempting to solicit attention; if jumping up on the owner ceases, jumping up has been negatively punished. The dog ventures too close to a boundary established with an electronic containment system and a warning tone sound. The dog runs back toward the house. This action turns off the unpleasant tone; if the dog stays away from the boundary in the future, running back home has been negatively reinforced (and moving near the boundary was positively punished).

Classical conditioning

Classical or respondent conditioning is another form of associative learning, meaning that associations or relations are established, but in this case, the relations are among stimuli. Events in the world often occur in predictable combinations and classical conditioning is what happens when an animal learns that one event reliably predicts another (Todes, 1997). This learning occurs because the animal finds the second event significant and in the presence of the first event, an animal comes to engage in behaviour that reflects the understanding that the second event will happen (Rescorla, 1988). The dog learns that when the owner opens a specific cupboard, a bowl of food will soon be presented. As a result, the dog begins to salivate when the cupboard door opens. The dog learns that being greeted by a stranger wearing a white lab coat predicts the insertion of a rectal thermometer or the prick of an injection. As a result, the dog begins to tremble and tries to escape the veterinary examination room. Respondent behaviour is often described as elicited, involuntary or reflexive; the animal is unable to control whether or not it salivates or trembles (Pierce and Cheney, 2004).

In the archetypal example of classical conditioning, Pavlov first established that placing meat powder in the mouth of a hungry dog elicits salivation (Pavlov, 1927). Prior learning was unnecessary for this to occur and, therefore, Pavlov termed the meat an unconditional stimulus (UCS) and salivation an unconditional response (UCR). Pavlov then preceded delivery of the meat powder with a distinctive sound, like a metronome. At the outset, the metronome produced only an orienting response. After a number of pairings of the metronome with the meat powder, the dog began to salivate upon hearing the metronome. Once the relation between the metronome and the meat powder was learned, the metronome became a conditional stimulus (CS) and the salivation elicited by the metronome, the conditional response (CR).

Social learning

Social or observational learning refers to any form of learning where changes in behaviour are dependent upon or facilitated by the presence of other animals. Thorpe (1963) proposes three categories of social learning: (i) social facilitation, in which the behaviour of one animal prompts the same behaviour by another animal but the behaviour already exists in the second animal's repertoire; (ii) local enhancement, in which the behaviour of one animal draws the attention of another animal to the specific place in the environment or the particular stimulus involved, but the second animal still acquires the new behaviour by trial and error; and (iii) true imitation, in which the behaviour of one animal prompts the exact same form of the behaviour by another animal and the behaviour is highly improbable in the second animal without this opportunity for observation.

Despite a lack of research on social learning in dogs, there is anecdotal evidence that dogs are capable of both social facilitation and local enhancement. An example of social facilitation is where a dog, reluctant to negotiate a set of slippery stairs, easily follows another dog up the stairs and, after that exposure, is comfortable traversing the stairs on its own. Trainers sometimes utilize social facilitation to encourage a dog to try a novel treat, to enter a new environment or to investigate a frightening object. An example of local enhancement is a dog that has never manipulated a food toy, like a Buster Cube™. After observing a demonstrator dog obtaining food from the toy, the second dog may learn a strategy for getting the food from the toy *more quickly* than a dog that was not given the opportunity to observe a demonstrator dog. Local enhancement is very likely the mechanism by which puppies in the Slabbert and Rasa (1997) study learned an odour-detection task more quickly after observing their dams performing the task than did control puppies that did not observe their dams at work.

True imitation can be concluded only after controlling for learning by social facilitation and local enhancement. Few, if any, convincing examples of imitation exist in species outside humans and the great apes (Whiten and Ham, 1992) and to my knowledge, there are no demonstrations of true imitative learning in dogs.

Establishing Desirable Behaviour through Operant Conditioning

One of the primary objectives of training the pet dog is to establish desirable behaviour, such as walking on a leash without pulling, lying down and remaining in place, and coming when called in the park. A more select subset of dogs are required to perform a repertoire of behaviours designed to fulfil specific functions, such as herding livestock, retrieving game, negotiating a course of agility obstacles, or assisting a physically compromised person. Dogs, like other species, show gradual acquisition of new behaviour mapping on to a negatively accelerating curve, such as in the hypothetical graph in Fig. 8.1. In this section, I outline the general processes by which dogs acquire instrumental behaviour for the purpose of attaining a pleasing outcome, in other words, learning by reinforcement.

Fig. 8.1. Hypothetical acquisition curve for dogs learning to come when called.

Motivation

In order to effect behavioural change through operant conditioning, there must be a motivating force in the form of a behavioural consequence (see also Chapter 4 in this volume). As noted above, behavioural consequences involve delivering or removing pleasant (appetitive) stimuli or delivering or removing unpleasant (aversive) stimuli. There are two classifications of such stimuli: unconditioned (or primary) and conditioned (secondary).

Unconditioned stimuli require no prior learning to appreciate, such as food, water, social contact, sex and various noxious or painful events. For ease of administration, most dog training utilizes food and social contact as forms of unconditioned appetitive reinforcers, with social contact consisting of vocalizations (praise), physical touch (stroking) and interactive play (wrestling, chase, tug and retrieving games). Not surprisingly, the value of these typical rewards tends to vary amongst dogs; some dogs find little reinforcing value in social contact, whereas other dogs prefer interactive play to food. One dog finds retrieving especially motivating, while another dog enjoys tug. A study in which dogs were required to press a lever for food rewards (see Fig. 8.2) revealed that one subject

Fig. 8.2. Photograph of dog bar-pressing in operant chamber.

would press the lever approximately 30–40 times per session for standard dry kibble rewards, but would press the same lever well over 100 times per session for small tidbits of liver (Reid, unpublished). In general, animals perform faster and/or longer for larger or more palatable rewards (Mackintosh, 1974).

Unconditioned aversive stimuli are usually vocalizations (reprimands), physical touch (restraint or other forms of 'corporal punishment' like hitting) or pain delivered through a training collar, such as choke or pinch collars that can be constricted around the dog's neck or electronic collars that can be triggered to present a noxious spray (citronella) under the dog's face or an electric shock to the dog's neck. Beerda *et al.* (1998) demonstrated that dogs show more fearful postures and experience elevated cortisol levels (indicative of higher stress) in response to unpredictable aversive events, such as loud noises, falling objects and electric shock, than to predictable events, like an umbrella opening, a spray of water from a gun or physical restraint. This is consistent with findings from controlled laboratory examinations of other species: the ability to predict an aversive event can dampen its unpleasantness somewhat (Lockard, 1963).

Operant conditioning is especially sensitive to the temporal relation between behaviour and its consequence (Mackintosh, 1974). In some situations, learning can be seriously hampered with as little as a 0.5 s delay between the behaviour and the outcome (Grice, 1948). In other cases, learning is possible with delays as long as several minutes (Lett, 1973). The critical feature appears to be the occurrence of competing behaviours during the delay (Revusky, 1971). For instance, if the dog sits but then barks, turns its head and jumps up, all before the trainer is able to deliver the food treat, it is difficult for the dog to discern which behaviour led to the reward. While the timely delivery of food is quite feasible in an experimental chamber, it is quite another matter in the real world of dog training, especially when the dog may be working at a distance from the trainer.

Fortunately, overcoming the problem of timing consequences is accomplished through the use of conditioned or secondary stimuli, which can be delivered immediately after the target behaviour. A conditioned reinforcer is a stimulus previously linked with an unconditioned reinforcer, such that it comes to predict the occurrence of the unconditioned reinforcement (Mackintosh, 1974). Pavlov's metronome functioned as conditioned reinforcement for his dogs. Practical conditioned reinforcers tend to be acoustic, so that they can be delivered immediately after the behaviour, even at a distance. For many dogs, 'Good dog!' is a conditioned reinforcer because it reliably precedes other pleasant events, like treats and play. 'Clicker training', first described by Skinner (1951) and subsequently popularized by marine mammal trainer Karen Pryor (1984), involves the use of a metal noisemaker, called a clicker, as a conditioned reinforcer (see Fig. 8.3). The dog is taught that the sound of the clicker signals a tangible reward is on its way and, consequently, the click bridges the temporal delay between behaviour and reward (Williams, 1991). A related but distinct effect is that the sound of the click helps to *mark* the target behaviour and make it distinguishable from other behaviours (Lieberman *et al.*, 1979); hence the clicker is often called a 'marker'. A trainer can time the administration of a click much more precisely than the

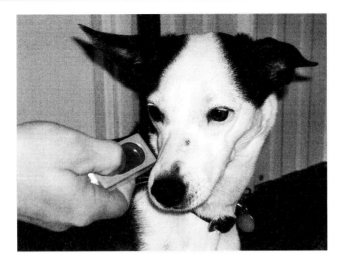

Fig. 8.3. Photograph of dog and clicker.

delivery of food treats or toys, thereby reducing uncertainty and speeding the learning process for the dog.

Less systematic in dog training is the use of conditioned negative reinforcement, also known as a safety signal. Imagine a situation where the dog jumps up on the kitchen counter. The owner shouts at the dog and the dog jumps off. Were the dog to remain, the owner would positively punish the dog in some way. Instead, the owner praises the dog for jumping back down. This praise functions as a safety signal to the dog, signalling that it has successfully avoided the positive punishment (I think of this as the 'Whew!' effect), thereby negatively reinforcing the action of jumping off the counter. Trainers who use devices like electronic collars sometimes establish tones as audible safety signals to convey that the dog has performed the target behaviour and escaped or avoided the potential aversive stimulus.

Acquiring the target behaviour

The simplest, and perhaps most overlooked, procedure for establishing desirable behaviour is to reinforce when the dog naturally emits the behaviour. Most responses that occur in the dog's behavioural repertoire can be increased in frequency beyond the baseline rate through reinforcement processes (responses that are not easily linked with the specific reinforcement being used are exceptions; see Shettleworth, 1972). Some trainers term this *catching* the behaviour. While effective, this technique is limited to naturally emitted behaviours.

In the laboratory environment, the primary acquisition procedure is a sequence of training steps called *shaping by successive approximations* (Skinner, 1951) or, more colloquially, free-form hand shaping. Initially the animal is reinforced for doing anything remotely similar to the target behaviour. Once an initial behaviour is established, reinforcement is systematically withheld so as to elicit variation in responding (behaviour is inherently variable and even more so when the animal becomes frustrated). Hopefully, a variation will occur that is closer in form to the target behaviour. This new behaviour is reinforced until established. Once again, reinforcement is withheld and the animal varies its behavioural responses. The trainer reinforces a new behaviour that is still closer to the target behaviour. These steps continue until the animal is performing the desired behaviour. For instance, suppose the goal is to teach the dog to take its leash off the hook by the door. Initially the dog is reinforced for looking in the direction of the door. Once the dog is performing this behaviour reliably, the trainer no longer reinforces this behaviour. The dog becomes frustrated and begins moving about the room. Eventually the dog moves in the direction of the door. The trainer reinforces any movement toward the door. Once the dog is reliably moving to the door area, the trainer no longer reinforces this behaviour. The dog hangs out by the door and eventually looks up at the hook. The trainer reinforces this behaviour. Once the dog is reliably going to the door and looking up at the hook, the trainer no longer reinforces this behaviour. In frustration, the dog reaches up toward the hook. The trainer reinforces this behaviour. The trainer continues to reinforce successive approximations to the target behaviour of the dog moving toward the door, reaching up to grasp the leash in its mouth, and pulling the leash from the hook. Shaping is still relatively new in the field of dog training, yet it is proving to be extremely useful for teaching trainers to objectively define the desired response, to hone timing and decision-making skills, and to remain patient while the dog 'offers' various responses.

Much more traditional in dog training is an acquisition procedure known as *prompting and fading* (Martin and Pear, 1996). Prompting involves introducing some stimulus event into the training situation so as to encourage the dog to perform the target behaviour. Taking the example from the previous paragraph, the dog could be prompted to perform the behaviour by leading the dog by the collar to the doorway and wiggling the leash to entice the dog to grab it and pull. The dog is then reinforced for performing the entire behaviour. This sequence is repeated, each time reducing or *fading* the amount of guidance provided until the dog is able to perform the behaviour on its own. Prompting can take the form of physical guidance or luring with an attractive item to elicit the behaviour from the dog so it can be reinforced.

In the case of training with negative reinforcement, often the primary purpose of the aversive stimulus is to prompt the desired behaviour to occur. One of the cleanest examples of negative reinforcement training with dogs is a technique for teaching the retrieve called the *ear pinch*. The trainer presents a retrieve object, such as a dumbbell, a few inches in front of the dog's muzzle and at the same time, pinches the dog's earflap to the extent that the dog finds it unpleasant. The pinch causes the dog to cry out in pain, at which point the trainer is able to insert the retrieve object into the dog's open mouth. At exactly that instant, the aversive stimulus (the pinch) is terminated, thus negatively reinforcing the dog for having the object in its mouth. Most dogs quickly learn to grab the dumbbell, consequently avoiding the painful pinch altogether. This behaviour can then be developed into a full retrieve, where the dog runs to collect the item and return it back to the trainer. Thus the ear pinch functions in two ways: (i) it prompts the dog to perform the desired behaviour of opening the mouth; and (ii) it reinforces grabbing the retrieve object through its termination. Readers, please bear in mind that I present this as an illustration of an acquisition procedure. Rest assured that there are equally effective, yet more humane methods for teaching a dog to retrieve objects.

More complex behaviours can be acquired through a procedure called *behaviour chaining* (Martin and Pear, 1996). Chaining involves breaking a complex behaviour down into smaller elements, establishing each of these separately, and then chaining the elements together to build the complete behaviour. Consider training a dog to place its toys in the toy box. This can be broken down into these elements: (i) the dog learns to pick a toy up from the floor; (ii) the dog learns to hold a toy in its mouth; (iii) the dog learns to walk while holding a toy in its mouth; (iv) the dog learns to walk from various locations to the toy box while holding a toy in its mouth; (v) the dog learns to hold its head over the toy box; and (vi) the dog learns to release the toy from its mouth. Each of these elements can be established separately and then chained together. Whenever possible, *backward chaining* is the preferred choice, meaning that the complex behaviour is established in reverse order (Mazur, 2002). The dog is first reinforced for dropping the toy in the box, then for carrying the toy to the box and dropping it in the box, then for picking up the toy, carrying it to the box and dropping it in, and finally, for moving to a toy on the floor, picking up the toy, carrying it to the box and dropping it in. This way the dog is always moving toward the terminal reinforcement. Each step along the way serves as conditioned reinforcement for the previous element. In contrast, *forward chaining* establishes the chain from the beginning. The dog is first reinforced for moving to the toy on the floor, then for moving to the toy on the floor and picking it up, and so forth. This procedure is less desirable because with each progression the dog is required to do more and so experiences a delay to the anticipated reinforcement.

Reinforcement schedules

A schedule of reinforcement refers to a rule that determines how and when a response will be followed by reinforcement (Ferster and Skinner, 1957). At the most basic level (i.e. simple schedules), reinforcers are delivered after a certain number of responses (a ratio schedule) or after the passage of a certain amount of time (interval or duration schedules). Researchers have collected a mass of data on how animals adjust and perform on a plethora of both simple and more complex reinforcement schedules (e.g. Morse, 1966) but, for the purposes of this chapter, I restrict the discussion to simple schedules. Different reinforcement schedules produce highly consistent patterns of responding. Traditionally, reinforcement schedules have been studied primarily in what is called *free operant* situations, in other words, the animal is free to perform the operant response as often as it likes and the measure of interest is the animal's rate of responding on a particular schedule. Dog training typically involves *discrete trials*, in which the dog can perform the operant response only during specified periods, such as when the trainer requests that the behaviour be performed. Consequently, rate of responding is not a relevant measure. Instead, schedule effects centre on the speed and reliability of the response.

It is well established that a continuous schedule of reinforcement (CRF) is optimum for the acquisition of new behaviour (Chance, 1999). CRF, also known as a *fixed ratio* 1 schedule (FR1), in which each response is reinforced, minimizes confusion as to what behaviour leads to reinforcement. Once the behaviour has been learned, however, it is beneficial to implement partial or intermittent schedules to maintain behaviour, in which responding is reinforced only some of the time. This benefit was initially discovered from observations of rats running alleyways to obtain food reinforcement. Surprisingly, rats ran faster when the goal box contained food only half the time (a fixed ratio 2) rather than all the time (Weinstock, 1958). Intermittent schedules have the valuable effect of teaching the dog to perform in the face of non-reward, a strength called *learned industriousness* (Eisenberger, 1992).

The larger the number of responses required for reinforcement, the more likely the animal will display a post-reinforcement pause (Myers and Mesker, 1960). Much like people procrastinating when starting a large task, animals take a break before responding on a large FR schedule. Trainers recognize that dogs willingly perform a large number of simple responses for a single reinforcer (for instance, a dog will run through a tunnel 25–100 times for one reinforcer) but may suffer from *ratio strain* when the trainer pushes for a large number of more complex responses (for instance, some dogs are unwilling to perform a sequence of more than 25 agility obstacles for a single reinforcer). Experienced trainers utilize progressive-ratio schedules, starting with a small requirement and gradually increasing the demand, to build up the dog's performance so that the potential for ratio strain is minimized.

Variable ratio (VR) schedules, in which the number of responses required for reinforcement varies in an unpredictable, albeit constrained manner, produce steady *high* rates of responding without a post-reinforcement pause (Ferster and Skinner, 1957). In general, responding on VR schedules is characterized by speedier, more intense performances. Those dog trainers who use intermittent schedules to maintain responding tend to rely on variable ratio schedules. For instance, after acquiring the behaviour of touching its nose to a target, a dog might be reinforced after two nose touches, then after four nose touches, then after one nose touch, then after five, and so forth. Again, the dog is prepared for this through a progressive-ratio schedule.

Occasionally, dog trainers incorporate reinforcement schedules that depend upon the passage of time. *Interval* schedules require that a certain amount of time must pass before the next response will be reinforced. Interval schedules can be fixed (a set amount of time – FI) or variable (the amount of time varies around a set mean – VI). Few examples of interval schedules exist in dog training. One notable exception is the trainer that works the dog off the leash and reinforces the dog for 'checking in' (the dog approaches and engages the attention of the trainer). Most trainers initially reinforce the dog for all 'check in' responses, but eventually the savvy trainer only reinforces a check-in that occurs some period of time *after* the last check-in (a VI schedule). This maintains check-ins at a steady, stable rate (Ferster and Skinner, 1957).

Often, dog trainers utilize time-based schedules that demand the animal sustain behaviour for a specific period of time. Martin and Pear (1996) refer to these schedules as *duration* schedules. For instance, dogs trained to stay, to maintain eye contact, and to move while in heel position, must do so for initially brief periods of time and gradually the amount of time required for reinforcement is extended. Like the other simple schedules, duration schedules can be fixed (always the same amount of time) or variable (the amount of time varies unpredictably).

Achieving stimulus control

Typically, an operant conditioning sequence consists of three events: (i) an antecedent stimulus; (ii) a response; and (iii) a consequential stimulus. For instance, the dog hears the trainer say 'lie down', the dog lies down, and the trainer delivers a food reward. Antecedent stimuli that precede operant behaviour and that *set the occasion* for responding are called discriminative stimuli (S^D) (Pierce and Cheney, 2004). A response that reliably occurs in the context of an S^D and not in the absence of the S^D is said to be under good stimulus control. Stimulus control is achieved through a process of differential reinforcement, such that the dog learns that responding in the presence of the S^D earns reinforcement, while responding in the absence of the S^D fails to produce reinforcement. In other words, the dog learns that the S^D sets the occasion for the response–reinforcer relation.

Trainers call achieving stimulus control as putting the behaviour *on cue*. The point at which this aspect of training is introduced depends upon the reliability of the response. If the response is highly probable, the S^D (or cue) can be introduced early in the acquisition phase because the trainer can be confident the response will occur in the presence of the S^D. If this is not the case, the trainer runs the risk of presenting the cue and the dog not responding. If this happens repeatedly, the cue is subject to *learned irrelevance* (Kremer, 1971), making it more difficult for the dog to learn that the cue sets the occasion for the response to be reinforced.

Whether a particular stimulus comes to control behaviour depends upon more than just arranging for it to precede responding. Animals are differentially sensitive to various stimuli. Dogs possess the predator's keen senses and are acutely aware of movement (Miller and Murphy, 1995). Suppose multiple stimuli are presented to the dog, such as when a trainer attempts to teach a dog to jump a hurdle on cue. The sight of the jump, the trainer's vocalization 'jump', and the trainer's movement toward the hurdle are all potential S^Ds. Given the dog's sensitivity to movement, the other stimuli are *overshadowed* and the dog's jumping behaviour comes to be controlled by the trainer's movement, rather than the trainer's presumed intention: the spoken word.

Indeed, it has been shown that animals are sometimes predisposed to link certain antecedent stimuli with certain behaviours. McConnell (1990) revealed that dogs are more likely to move in the presence of high-pitched, rapidly repeating sounds, like 'come, come, come!' and are more likely to remain stationary in the presence of low-pitched, continuous sounds, like 'down'. In fact, McConnell and Bayliss (1985) determined that across a wide range of cultures and languages, the cues we use to induce action from dogs tend to be high-pitched, rapidly repeating sounds and, in contrast, the cues we use to inhibit movement in dogs are low-pitched, continuous sounds. The selectivity of stimulus and response is especially pervasive when it comes to learning by negative reinforcement. Studies with numerous species demonstrate speedy learning when the required avoidance response is consistent with the species' defence responses, and painfully slow when the required response is inconsistent (Bolles, 1970). For instance, a pigeon easily associates sounds, but not sights, with aversive stimuli and quickly learns to flap its wings in order to avoid the stimulus. The pigeon would have a difficult time learning a response that is inconsistent with defence, like pecking a key, to avoid an aversive stimulus (Mazur, 2002).

Pavlov noticed that after one stimulus, like the metronome, was conditioned to the point that it elicited salivation, other similar stimuli would also elicit salivation. This is called *stimulus generalization*. The extent to which generalization across stimuli occurs can be portrayed in a stimulus generalization gradient. Figure 8.4 shows a gradient of responding as a function of how similar the stimuli are to the original test stimuli. These data are hypothetical. The dog was trained to sit when it hears the word 'sit'. However, it also shows some probability of responding with

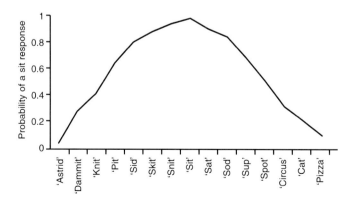

Fig. 8.4. Hypothetical generalization gradient for a dog responding to verbal cues.

a sit when it hears similar words like 'hit', 'spit' and 'lit'. It is less likely to respond to words like 'spot' and 'cat' and even less likely with highly dissimilar words like 'pizza' and 'envelope'.

The steepness of a stimulus gradient is a measure of how sensitive the animal is to variations in a particular class of stimuli. A very flat gradient indicates that the dog responds to a wide range of stimuli, in other words, the dog is under poor stimulus control. A very steep gradient indicates that the dog responds very selectively to a stimulus, in other words, the dog is under good stimulus control. Steep gradients are usually produced through differential reinforcement. My dog 'Eejit' reliably responded to his name, but when I adopted 'Fidget', not surprisingly, he initially responded to her name as well. It took about one week for him to learn to differentiate his name (the S^+, the stimulus that signals significant events for him) from hers (the S^-, the stimulus that signals nothing for him). This is the essence of simple discrimination training. The traditional way of teaching a simple discrimination through differential reinforcement is a slow process and the animal makes many errors. Imagine how many times Eejit responded to the name 'Fidget' and was ignored before he ceased responding. Often this type of training is unpleasant for the animal. Terrace (1963) developed a superior method called *errorless discrimination learning*, errorless because the animal makes few or no responses to the S^-. Initially, the S^+ is made attractive to the animal so it is highly likely to respond to it. At the same time, the S^- is faded in so that the animal is unlikely to respond to it. A realistic example in dog training is a task called scent discrimination. The dog is required to search a set of visually similar items, one of which also contains the scent of the trainer. The dog is to retrieve the item that smells of the trainer (the S^+) and ignores the other items (the S^-). Suppose initially the S^+ item is also impregnated with the smell of food, like liver. At the same time, the S^- items are presented, at first only one but gradually more and more. The S^- items are smaller than the S^+ and placed far away from the dog. The dog is highly likely to select the correct item. Over trials, the S^- items are made progressively more apparent, by increasing their size and by placing them closer and closer to the S^+ item. But the dog has been repeatedly rewarded for selecting the S^+ item so, by the time the S^- items are salient, it is improbable that they will elicit responding. In one of Terrace's studies (1963), pigeons trained with a conventional discrimination procedure made more than 3000 responses to the S^- during 28 sessions. In contrast, pigeons trained with errorless discrimination learning made about 25 responses to the S^- during a comparable number of sessions. Terrace's procedure produces reliable discrimination learning in a minimal number of training sessions.

Dog trainers will attest that a much more common concern is that dogs discriminate too well; that they *fail* to generalize. A comprehensive training programme should include generalization training, also known as *stimulus equivalence training*. We often expect dogs to generalize across a wide range of stimuli, such as responding to the word 'sit' spoken by men, women and children, people both familiar and unfamiliar to the dog. It appears that dogs are unlikely to generalize this widely without specific training in which the cue 'sit' is presented first in the original form and then in forms gradually less similar until the dog responds to the common stimulus 'sit' in virtually any form. It turns out that responding also comes under the control of cues provided by the context in which the original training took place (Stokes and Baer, 1977). This is the origin for the common complaint of pet owners, 'but he comes when he's called in the yard but not at the park'. The original training was conducted in the yard, with distinct contextual cues and few stimuli to compete for the dog's attention. Lacking targeted equivalence training, the dog is unlikely to generalize coming when called to the park setting, with very dissimilar contextual cues and complete with a multitude of stimuli competing for the dog's attention.

Suppressing Unwanted Behaviour through Operant Conditioning

A second primary objective for training dogs is to reduce or suppress the frequency of undesirable behaviour. Pet dog owners find themselves faced with a myriad of normal, yet objectionable dog behaviours: dogs bark at inappropriate times, urinate on the furniture, pull on the leash, barge out the front door and growl at strangers. Working dogs also display behaviours that interfere with successful performance of their tasks. Herding dogs bite their charges; retrieving dogs run off with the game; agility dogs are so keen that they leave obstacles prematurely; and guide dogs lead their owners too close to moving traffic. Some type of punishment, whether positive or negative, may be required to eliminate such problematic behaviour.

In this section, I outline the general processes by which dogs inhibit instrumental behaviour for the purpose of escaping or avoiding an unpleasant outcome, in other words, learning by punishment. The stimuli used to motivate such learning are the same as for learning by reinforcement: appetitive and aversive stimuli. However, negative punishment can result from the cessation or prevention of appetitive stimuli; and positive punishment can result from the threat or presentation of aversive stimuli. I also include a procedure called extinction that is not punishment but still functions to produce a decline in behaviour.

Extinction

Extinction involves the discontinuation of reinforcement for a conditioned behaviour. As such, any associative learning, whether operant or classical, can be subject to extinction. However, the most common application of extinction is to cease reinforcement, whether intentional or adventitious, of established operant behaviour. For instance, imagine the pet dog owner who, quite innocently, teaches the dog to whine to be released from confinement. Initially, the dog whines from the distress of separation. The owner releases the dog and whining is negatively reinforced with escape from an unpleasant situation. Eventually the dog learns to whine relentlessly whenever it is placed in confinement until the owner breaks down and releases it. If the owner commits to ignoring the dog when it whines, in other words, omitting the reinforcement, whining will extinguish.

The goal of extinction is to reverse the effects of acquisition. Extinction does indeed lead to a gradual decline in the frequency of a conditioned behaviour. The confined dog gradually, over time, whines less. However, the first effect of extinction is that the behaviour becomes more variable (Neuringer *et al.*, 2001). Maybe the dog tries barking, scratching or digging. Nonreinforcement produces frustration and frustration typically energizes behaviour (Amsel, 1962). Whining may actually increase beyond previous levels, called an *extinction burst*. Frustration also produces strong emotional responses, such as aggression (Azrin *et al.*, 1966). If a second animal is present, extinction-induced aggression may be directed toward the innocent third party or toward the owner.

One of the hallmark effects of extinction is *spontaneous recovery* (Rescorla, 1997). Suppose the dog ceases whining and is released from confinement. The next time the dog is placed in confinement, whining resumes for a while. Recovery of the response is rarely complete and extinction occurs more rapidly than the first time. With each subsequent period of confinement, the extent of spontaneous recovery diminishes. The occurrence of spontaneous recovery can be devastating to a dog owner. Other recovery effects can also cause a restoration of the unwanted behaviour. Confine the dog in a different environment and whining is likely to emerge; this is called *renewal* of the response. Renewal occurs because extinction was linked to the contextual cues present in the original situation (Bouton and King, 1983). Exposing the dog to the reward of being released from confinement, or even the stimuli leading up to the reward, such as attending to the dog, can also prompt resumption of whining. This is called *reinstatement* (Bouton, 1994).

Yet another potential limitation of extinction is the *partial reinforcement extinction effect*. Prior to extinction, the conditioned behaviour was reinforced on either a continuous or an intermittent schedule. Behaviours with an intermittent reinforcement history extinguish much more slowly than behaviours with a continuous reinforcement history (Jenkins and Stanley, 1950). Think of how long the string of losses would need to be for a habitual gambler to stop playing the slot machines! Extinction that follows partial reinforcement also produces fewer frustration responses. Resistance to extinction can also be beneficial. Desirable behaviours that have been maintained on partial reinforcement are likewise resistant to extinction. Dogs can be expected to perform reliably in the competition ring where tangible reinforcement is prohibited, if the training incorporated intermittent reinforcement.

Instituting extinction is a popular procedure for reducing unwanted behaviour because it does not involve the use of aversive stimuli. However, there are so many variables prompting restoration of the behaviour that the practical value of extinction is extremely limited.

Negative punishment

Negative punishment can be used to discourage or suppress unwanted behaviour in cases where the animal is anticipating a pleasant outcome. When the unwanted behaviour occurs, delivery of the appetitive stimulus is prevented or, if the appetitive stimulus has already been presented, occurrence of the unwanted behaviour causes withdrawal of the appetitive stimulus. Suppose an owner wants the dog to cease jumping up to greet them upon their arrival home after an absence. The dog jumps up in anticipation of a friendly reunion. If, instead, the owner turns away and ignores the dog for a period of time, the anticipated reward is withheld. This is an application of negative punishment and one would expect the likelihood of jumping up to diminish.

Negative punishment sounds a lot like extinction but it is distinctly different. Extinction occurs when a previously reinforced behaviour no longer produces reinforcement. In the case of negative punishment, there is no requirement that the behaviour be reinforced previously. The essence of negative punishment is that when the unwanted behaviour occurs, sources of reinforcement are withheld for some period of time. In most applications of negative punishment, it is commonly referred to as *time out*. In some cases, only the existing reinforcement is withheld; in other cases, all sources of reinforcement are removed. For instance, arranging for the dog owner to turn away from the dog when it jumps up prevents the dog from enjoying a social greeting. However, removing the dog to a small area of confinement, with no access to alternative activities, temporarily restricts the dog from all potential sources of reinforcement. While we might assume that removal from all sources would be more effective, sometimes it is difficult to achieve because the act of moving the dog to an isolation area can be problematic and, once there, dogs are adept at finding ways to entertain themselves with other undesirable behaviour (chewing, digging, barking, etc.). In general, short time-out durations of a few seconds to a minute or two are effective (Lindsay, 2000). Longer time out periods do not increase effectiveness (White *et al.*, 1972).

The use of time out to eliminate unwanted behaviour during controlled situations, such as training sessions, can be extremely effective, provided the dog finds training pleasurable. For instance, a savvy trainer ignores the dog for a second or two when the dog barks rather than performing the requested behaviour, another trainer requires that the agility dog lie down for a moment after performing an obstacle incorrectly, and still another trainer sends the dog into its crate after an infraction so that it must wait while the trainer works another dog. However, it should be clear that these examples of time out would not be appropriate for a dog that would prefer to avoid training!

Like extinction, time out is an appealing technique for reducing unwanted behaviour because it does not involve the use of aversive stimuli. A few studies have contrasted the effectiveness of extinction and negative punishment with positive punishment, suggesting that: (i) negative punishment is more effective than extinction in suppressing unwanted behaviour; and (ii) negative punishment takes longer than positive punishment to reduce the frequency of unwanted behaviour and that suppression is rarely complete (Uhl and Sherman, 1971).

Positive punishment

Positive punishment is the presentation of an aversive stimulus, contingent upon the occurrence of an unwanted behaviour. If the procedure is effective, the unwanted behaviour should decrease in frequency or be eliminated altogether. For example, suppose an owner wants to discourage the dog from getting up on the couch. The owner could watch the dog carefully and each and every time the dog jumps on the couch, the owner loudly shakes an aluminium can filled with pebbles. After a few attempts, the dog learns that the way to avoid hearing the noxious sound is to stay clear of the couch. Learning curves are often steeper with positive punishment than with other operant conditioning procedures. Positive punishment can produce dramatic suppression of behaviour in just a few repetitions (Domjan, 2003).

Much of the research on positive punishment has been devoted to delineating its use. In general, positive punishment leads to more effective suppression of behaviour when these conditions are met:

- The more intense and/or long lasting the aversive stimulus, the more behavioural suppression (low intensity aversive stimulation produces only moderate suppression and eventual response recovery) (Azrin, 1960).
- An aversive stimulus introduced at high intensity is more effective than if initially presented at low intensity and then increased (animals become resistant to the effects of the aversive stimulus if it is initially presented at low intensity; conversely experience with high intensity aversive stimulation increase the effectiveness of later low intensity aversive stimuli) (Azrin and Holz, 1966).
- Even response-independent positive punishment can lead to some suppression, but response-dependent positive punishment is significantly more effective (in other words, exposure to an aversive stimulus, whether contingent on behaviour or not, leads to some dampening down of *all* instrumental behaviour) (Church *et al.*, 1970).
- Immediate punishment (both negative and positive) is more effective than delayed punishment (Camp *et al.*, 1967).
- The degree of suppression produced by positive punishment is a direct function of the number of responses punished (intermittent punishment can work, but not as effectively as a continuous schedule of punishment) (Azrin, 1956).

Positive punishment is more effective when the animal is provided with other appropriate activities for obtaining reinforcement (Herman and Azrin, 1964). For instance, a dog jumps on the couch for the reinforcement of a soft place to lie down. If the dog is provided with its own comfortable resting area, punishing the dog for getting on the couch is more likely to work than if the dog has no alternative for obtaining the reinforcement. If a discriminative stimulus signals that positive punishment is in effect, the animal will learn to suppress behaviour only in the presence of that S^D (Dinsmoor, 1952). If the dog is only punished for getting on the couch when the owner is home, the dog will only learn to refrain from getting on the couch when the owner is present.

The use of positive punishment (and negative reinforcement), because they entail aversive stimuli, is highly controversial. This issue is not whether positive punishment works; there is no question that the aversive control of behaviour is effective. The concern is whether the use of positive punishment is ethical, especially for animals, because they cannot consent. When deciding whether the benefits of positive punishment warrant its use, there are numerous factors to be considered, including the expertise of the trainer, the seriousness of the unwanted behaviour, the urgency of the need for behaviour change, and the quality of life of the dog and of the humans coming into contact with the dog. Another concern is that the line between positive punishment and abuse can be unintentionally crossed, both by pet owners and skilled professionals. Indeed, dog-training texts dating pre-1990s are peppered liberally with techniques that waver along that line, such as scruff shakes, alpha rolls, severe collar corrections, hanging and helicoptering (e.g. Saunders, 1946; Koehler, 1962; Strickland, 1965; Monks of New Skeet, 1978; Brown, 1983). Fortunately, as dog trainers become well versed in learning theory, more humane applications are being developed. While it is outside the scope of this chapter to explore the controversy in detail, let me suggest that animal care professionals comply with the principle of the *least aversive alternative*: that the least aversive technique that is reasonably expected to succeed should be the technique of choice (Delta Society, 2001). Adherence to this principle will at least give pause for thought and, hopefully, people will err on the side of humane treatment.

Behaviour Change through Classical Conditioning

Pavlov's pioneering work on how dogs come to anticipate the presentation of food has proliferated into an extensive body of research exploring the formation of associations between stimuli (Rescorla, 1988). A major focus has been the conditioning of emotional reactions such as fear. Watson and Raynor (1920) demonstrated the ease with which a child can learn to fear an object. 'Albert', a 9-month-old boy, initially showed no fear of a white rat, in fact he wanted to touch it, but he did show distinct displeasure upon hearing the loud sound of a hammer hitting a steel bar. The researchers then went about presenting the rat, followed by the sound. After just two repetitions, Albert was reluctant to touch the rat. With just a few more repetitions, Albert cried and tried to move away whenever he saw the rat. This is a classic example of a 'conditioned emotional response'.

Emotional responses, such as excitement, anxiety, fear and aggression, are often triggered by specific stimuli. Envision the dog that gets wildly excited when its owner picks up the leash, anticipating a walk in the park. Or the dog that begins to pace and whine when its owner prepares to leave for the workday, anxious at the prospect of being left alone. Or the dog that hides under the bed when the owner picks up the nail clippers, fearing a painful grooming episode. Conditioned emotional responses not only underlie many serious behavioural problems, they can also hinder the learning and performance of desirable behaviour. A pet dog may be too anxious to benefit from training classes or a working dog may be so aggressive that it can't be trusted to perform in the required circumstances.

Classical conditioning procedures geared toward reducing interfering emotional responses are typically more effective at remedying problems such as fear and aggression than operant conditioning procedures geared toward addressing the behaviour directly. For instance, suppose you have a dog that barks and lunges aggressively at people entering your home. One option is to train the dog to maintain a sit-stay and to suppress barking, presumably through negative or positive punishment, when people enter the home. This is an operant conditioning approach and it would be effective in altering the dog's behaviour, but not necessarily the dog's underlying emotional state (Wright *et al.*, 2005). A second option is to teach the dog a pleasant association: each time a person enters the home, wonderful things happen to the dog. Maybe the visitor always brings a large meaty bone for the dog. With sufficient repetitions, the dog comes to anticipate bones when people come to the door. The dog now experiences a new kind of conditioned emotional response, one of pleasure and excitement, rather than anxiety and aggression. One would predict the dog's behaviour to change from lunging and barking to opening the door wide and welcoming guests to your home!

Desensitization and counterconditioning

The most popular approach for eliminating problematic behaviours, such as fearful and aggressive responses, is *desensitization and classical counterconditioning* (DSCC). Wolpe (1958) was the first to describe the use of DSCC to eradicate conditioned fear in cats. He taught hungry cats to fear the presence of food by associating the act of feeding with the delivery of electric shock. After the initial conditioning, the cats refused to eat in the experimental room, despite significant food deprivation. The cats also generalized to rooms that were similar to the experimental room. Wolpe identified a room that was sufficiently unlike the experimental room that the cats, though still anxious, were able to eat. Successive feedings in the new room eliminated all signs of distress. The cats were then moved to a room slightly more similar in appearance to the experimental room and offered food. This was repeated in a series of rooms of increasing similarity to the experimental room, each time remaining in the same room until distress was no longer visible. Eventually the cats were able to eat comfortably in the original room.

This same technique can be applied to a fearful dog, such as that shown in Fig. 8.5. This dog became tense when people reached for it. If escape were possible, it would retreat. If not, it would snap and bite. I started by pairing a hand lifted in the air, but not directed toward the dog, with highly palatable treats. Once the dog was comfortable with this arrangement, I progressed to orienting the hand toward the dog. Gradually the hand was moved progressively closer to

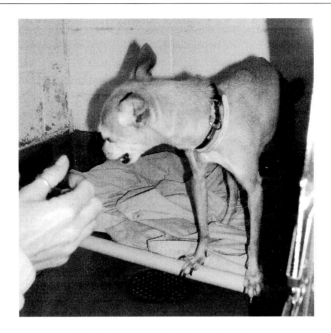

Fig. 8.5. Photograph of chihuahua recoiling from a hand.

the dog, remaining at each step until the dog appeared to relax. Eventually, the dog appeared eager to have a hand reaching out to touch it, even when treats weren't offered immediately (see Fig. 8.6). Because the dog no longer feared being touched, defensive behaviours were replaced with approach and tail wagging.

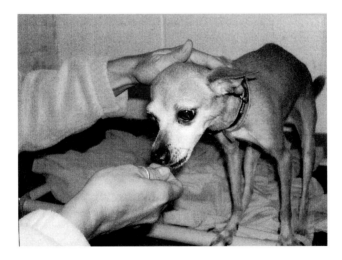

Fig. 8.6. Photograph of chihuahua tolerating hand touch.

Classical counterconditioning, in particular, is an extremely powerful agent for behaviour change. In most laboratory studies of counterconditioning (also known as cross-motivational transfer experiments), an initially neutral stimulus is first paired with one unconditioned stimulus, such as shock, and then, at a later time, the same previously neutral stimulus is paired with a motivationally disparate stimulus, such as food (Dickinson and Dearing, 1979). In even more convincing demonstrations, though, two incompatible unconditioned stimuli are paired together. Erofeeva (1921, cited in Dickinson and Dearing, 1979) used a strong electric shock to signal the delivery of food to hungry dogs. The dogs initially responded with defensive behaviours, such as struggling and yelping. Yet, as conditioning progressed, the dogs began to show typical appetitive responses in response to the shock, including lip licking and salivation. Even more surprising, after counterconditioning the aversive stimulus to signal an appetitive stimulus, the aversive stimulus was incapable of functioning as a punishing stimulus in an operant conditioning procedure. Classical counterconditioning is so potent that it appears to actually produce a change in the motivational and reinforcing properties of an unconditioned stimulus.

Concluding Remarks

Learning theory is one of the most in-depth and comprehensive accounts of behaviour to appear in the psychological literature. It has generated countless testable predictions, spawned entire research careers, and sprouted highly successful applications in the clinical world of human and animal behaviour. Learning theory provides an expansive and thorough framework for understanding the processes by which dogs learn and how best to alter their behaviour. Such knowledge is proving to spark the development of more effective, efficient and humane methods of dog training and behaviour modification.

References

Amsel, A. (1962) Frustrative nonreward in partial reinforcement and discrimination learning. *Psychological Review* 69, 306–328.
Azrin, N.H. (1956) Some effects of two intermittent schedules of immediate and non-immediate punishment. *Journal of Psychology* 42, 3–21.
Azrin, N.H. (1960) Effects of punishment intensity during variable-interval reinforcement. *Journal of the Experimental Analysis of Behavior* 3, 123–142.
Azrin, N.H. and Holz, W.C. (1966) Punishment. In: Honig, W.K. (ed.) *Operant Behavior: Areas of Research and Application.* Appleton-Century-Crofts, New York, pp. 380–447.
Azrin, N.H., Hutchinson, R.R. and Hake, D.F. (1966) Extinction-induced aggression. *Journal of the Experimental Analysis of Behavior* 9, 191–204.
Beerda, B., Schilder, M.B.H., van Hooff, J.A., de Vries, H.W. and Mol, J.A. (1998) Behavioural, saliva cortisol and heart rate responses to different types of stimuli in dogs. *Applied Animal Behaviour Science* 58, 365–381.

Bolles, R.C. (1970) Species-specific defense reactions and avoidance learning. *Psychological Review* 71, 32–48.

Bouton, M.E. (1994) Conditioning, remembering, and forgetting. *Journal of Experimental Psychology: Animal Behavior Processes* 20, 219–231.

Bouton, M.E. and King, D.A. (1983) Contextual control of the extinction of conditioned fear: tests for the associative value of the context. *Journal of Experimental Psychology: Animal Behavior Processes* 9, 248–265.

Brown, B.G. (1983) *The No-force Method of Dog Training. Book One: Novice.* H and S Publications, Galesburg, Illinois.

Camp, D.S., Raymond, G.A. and Church, R.M. (1967) Temporal relationship between response and punishment. *Journal of Experimental Psychology* 74, 114–123.

Chance, P. (1999) *Learning and Behavior*, 4th edn. Brooks/Cole, Pacific Grove, California.

Church, R.M., Wooten, C.L. and Matthews, T.J. (1970) Discriminative punishment and the conditioned emotional response. *Learning and Motivation* 1, 1–17.

Coppinger, R. and Coppinger, L. (2001) *Dogs: A Startling New Understanding of Canine Origin, Behavior, and Evolution.* Scribner, New York.

Davis, M. (1974) Sensitization of the rat startle response by noise. *Journal of Comparative and Physiological Psychology* 87, 571–581.

Delta Society (2001) *Professional Standards for Dog Trainers: Effective, Humane Principles.* Delta Society, Bellevue, Washington.

Dickinson, A. and Dearing, M.F. (1979) Appetitive-aversive interactions and inhibitory processes. In: Dickinson, A. and Boakes, R.A. (eds) *Mechanisms of Learning and Motivation: A Memorial to Jerzy Konorski.* Lawrence Erlbaum Associates, Hillsdale, New Jersey, pp. 203–231.

Dinsmoor, J.A. (1952) A discrimination based on punishment. *Quarterly Journal of Experimental Psychology* 4, 27–45.

Domjan, M. (2003) *The Principles of Learning and Behavior*, 5th edn. Wadsworth/Thomson, Belmont, California.

Eisenberger, R. (1992) Learned industriousness. *Psychological Review* 99, 248–267.

Ferster, C.B. and Skinner, B.F. (1957) *Schedules of Reinforcement.* Appleton-Century-Crofts, New York.

Grice, G.R. (1948) The relation of secondary reinforcement to delayed reward in visual discrimination learning. *Journal of Experimental Psychology* 38, 1–16.

Herman, R.L. and Azrin, N.H. (1964) Punishment by noise in an alternative response situation. *Journal of the Experimental Analysis of Behavior* 7, 185–188.

Jenkins, W.O. and Stanley, J.C. (1950) Partial reinforcement: a review and critique. *Psychological Bulletin* 47, 193–234.

Koehler, W.R. (1962) *The Koehler Method of Dog Training.* Howell Book House, New York.

Kremer, E.F. (1971) Truly random and traditional control procedures in CER conditioning in the rat. *Journal of Comparative and Physiological Psychology* 76, 441–448.

Leaton, R.N. (1976) Long-term retention of the habituation of lick suppression and startle response produced by a single auditory stimulus. *Journal of Experimental Psychology: Animal Behavior Processes* 2, 248–259.

Lett, B.T. (1973) Delayed reward learning: disproof of the traditional theory. *Learning and Motivation* 3, 237–246.

Lieberman, D.A., McIntosh, D.C. and Thomas, G.V. (1979) Learning when reward is delayed: a marking hypothesis. *Journal of Experimental Psychology: Animal Behavior Processes* 5, 224–242.

Lindsay, S.R. (2000) *Handbook of Applied Dog Behavior and Training: Vol. 1. Adaptation and Learning.* Iowa State University Press, Ames, Iowa.

Lockard, J.S. (1963) Choice of a warning signal or no warning signal in an unavoidable shock situation. *Journal of Comparative and Physiological Psychology* 56, 526–530.

Mackintosh, N.J. (1974) *The Psychology of Animal Learning.* Academic Press, London.

Martin, G. and Pear, J. (1996) *Behavior Modification: What It Is and How To Do It*, 5th edn. Prentice Hall, Upper Saddle River, New Jersey.

Mazur, J.E. (2002) *Learning and Behavior*, 5th edn. Prentice Hall, Upper Saddle River, New Jersey.

McConnell, P.A. (1990) Acoustic structure and receiver response in domestic dogs, *Canis familiaris. Animal Behaviour* 39, 897–904.

McConnell, P.A. and Bayliss, J.R. (1985) Interspecific communication in cooperative herding: acoustic and visual signals from human shepherds and herding dogs. *Zeitschrift für Tierpsychologie* 67, 303–328.

Miller, V. and Domjan, M. (1981) Selective sensitization induced by lithium malaise and footshock in rats. *Behavioral and Neural Biology* 31, 42–55.

Miller, P.E. and Murphy, C.J. (1995) Vision in dogs. *Journal of the American Veterinary Medical Association* 207, 1623–1634.

Monks of New Skeet (1978) *How To Be Your Dog's Best Friend.* Little, Brown, Boston, Massachusetts.

Morse, W.H. (1966) Intermittent reinforcement. In: Honig, W.K. (ed.) *Operant Behavior: Areas of Research and Application.* Appleton-Century-Crofts, New York, pp. 52–108.

Myers, R.D. and Mesker, D.C. (1960) Operant conditioning of a horse under several schedules of reinforcement. *Journal of the Experimental Analysis of Behavior* 3, 161–164.

Neuringer, A., Kornell, N. and Olufs, M. (2001) Stability and variability in extinction. *Journal of Experimental Psychology: Animal Behavior Processes* 27, 79–94.

Pavlov, I.P. (1927) *Conditioned Reflexes* (G.V. Anrep, translation). Oxford University Press, London.

Pierce, W.D. and Cheney, C.D. (2004) *Behavior Analysis and Learning.* Lawrence Erlbaum Associates, Mahwah, New Jersey.

Pryor, K. (1984) *Don't Shoot the Dog: The New Art of Teaching and Training.* Bantam Books, New York.

Rescorla, R.A. (1988) Pavlovian conditioning: it's not what you think it is. *American Psychologist* 43, 151–160.

Rescorla, R.A. (1997) Spontaneous recovery after Pavlovian conditioning with multiple outcomes. *Animal Learning and Behavior* 25, 99–107.

Revusky, S. (1971) The role of interference in association over a delay. In: Honig, W.K. and James, P.H.R. (eds) *Animal Memory.* Academic Press, New York, pp. 155–213.

Salzinger, K. and Waller, M.B. (1962) The operant control of vocalization in the dog. *Journal of the Experimental Analysis of Behavior* 5, 383–389.

Saunders, B. (1946) *Training You to Train Your Dog.* Doubleday, New York.

Shettleworth, S.J. (1972) Constraints on learning. *Advances in the Study of Behavior* 4, 1–68.

Skinner, B.F. (1938) *The Behavior of Organisms.* Appleton-Century-Crofts, New York.

Skinner, B.F. (1951) How to teach an animal. *Scientific American* 185, 26–29.

Slabbert, J.M. and Rasa, O.E.A. (1997) Observational learning of an acquired maternal behaviour pattern by working dog pups: an alternative training method? *Applied Animal Behaviour Science* 53, 309–316.

Stokes, T.F. and Baer, D.M. (1977) An implicit technology of generalization. *Journal of Applied Behavior Analysis* 10, 349–367.

Strickland, W.G. (1965) *Expert Obedience Training for Dogs.* Macmillan, New York.

Terrace, H.S. (1963) Discrimination learning with and without 'errors'. *Journal of the Experimental Analysis of Behavior* 6, 1–27.

Thompson, R.F. and Spencer, W.A. (1966) Habituation: a model phenomenon for the study of neuronal substrates of behavior. *Psychological Review* 73, 16–43.

Thorndike, E.L. (1911) *Animal Intelligence.* Macmillan, New York.

Thorpe, W.H. (1963) *Learning and Instinct in Animals.* Harvard University Press, Cambridge, Massachusetts.

Todes, D.P. (1997) From the machine to the ghost within: Pavlov's transition from digestive physiology to conditioned reflexes. *American Psychologist* 52, 947–955.

Uhl, C.N. and Sherman, W.O. (1971) Comparison of combinations of omission, punishment, and extinction methods in response elimination in rats. *Journal of Comparative and Physiological Psychology* 74, 59–65.

Watson, J.B. and Raynor, R. (1920) Conditioned emotional reactions. *Journal of Experimental Psychology* 3, 1–14. Reprinted in *American Psychologist* 55, 313–317.

Weinstock, S. (1958) Acquisition and extinction of a partially reinforced running response at a 24-hour intertrial interval. *Journal of Experimental Psychology* 46, 151–158.

White, G.D., Nielsen, G. and Johnson, S.M. (1972) Timeout duration and the suppression of deviant behavior in children. *Journal of Applied Behavior Analysis* 5, 111–120.

Whiten, A. and Ham, R. (1992) On the nature and evolution of imitation in the animal kingdom: reappraisal of a century of research. *Advances in the Study of Behavior* 21, 239–283.

Williams, B.A. (1991) Marking and bridging versus conditioned reinforcement. *Animal Learning and Behavior* 19, 264–269.

Wolpe, J. (1958) *Psychotherapy by Reciprocal Inhibition.* Stanford University Press, Stanford, California.

Wright, J.C., Reid, P.J. and Rozier, Z. (2005) Treatment of emotional distress and disorders – non-pharmacological methods. In: McMillan, F.D. (ed.) *Mental Health and Well-being in Animals.* Blackwell Publishing, Ames, Iowa, pp. 145–157.

III The Dog in Its Niche: Among Humans

Editor's Introduction

The ecological concept of a 'niche' is used in many places throughout this book. In normal biological texts, a niche is a specific living environment, signified by all specific aspects of how a population utilizes its living space, such as the food eaten, the geographical place occupied, etc. It has become increasingly obvious that for dogs the most fundamental aspect of the niche it occupies is ourselves. One might therefore say that dogs have come to occupy the ecological niche of living with humans. The third part of the book explores the behavioural biology of dogs from this perspective.

A first glimpse of how dogs have adapted to this niche is offered in Chapter 9, which deals with the behaviour of feral dogs. These are populations of dogs which have more or less escaped from the most intense life with humans, in that they are not living in families. The traits of such dogs therefore offer many insights into how their behaviour has been modified since they roamed the continents as wild wolves. Chapter 10 introduces the problem of selecting and breeding dogs for various working purposes. Here the author provides an overview of some important aspects of modern population genetics and applies this to dog breeding and selection.

Any selection and breeding obviously depends on how well we can characterize the behaviour we intend to select for. A much overlooked aspect of dog behaviour is the concept of personality, and this forms the central part of Chapter 11. The concept raises fundamental issues with respect to its definition, but also of course necessitates careful consideration when one wishes to measure personalities. The last chapter of this part, Chapter 12, deals with the way in which dogs and humans have co-evolved to an extent not paralleled by any other domestic species. The author provides a number of intriguing research results concerning the social cognitive abilities of dogs, and also some unexpected aspects of human competence in communicating with dogs.

Behaviour and Social Ecology of Free-ranging Dogs

Luigi Boitani, Paolo Ciucci and Alessia Ortolani

There Are Many Kinds of Free-ranging Dogs

More than 350 officially recognized dog breeds and the endless variety of mongrels are the best evidence of the broad morphological adaptability of dogs to different environments, functions and human-driven selection. Similar plasticity, although not equally evident, can be found in dogs' behaviour and ecology when they are exposed to the variety of artificial and natural selections in almost any possible environment on Earth. Because of this diversity of phenotypes and functional specializations, dogs are a highly diverse biological group that allows only limited and superficial generalizations. In particular, when studying dog behaviour, it is imperative to distinguish at least a few broad dog categories defined by their social context and their individual behavioural ontogenies. Dogs have been classified on the basis of: behavioural and ecological traits (Scott and Causey, 1973; Causey and Cude, 1980); their origins (Daniels and Bekoff, 1989a, b); their main range type (rural vs urban free-ranging: Berman and Dunbar, 1983; those having unrestricted access to public property: Beck, 1973); and the extent of their dependency on humans (World Health Organization (WHO), 1988). The latter classification was designed to improve dog rabies control measures (Perry, 1993) and is useful to define dogs in relation to their association with humans: (i) restricted dog: fully dependent (all its essential needs provided intentionally by humans) and fully restricted by humans; (ii) family dog: fully dependent and semi-restricted; (iii) neighbourhood dog: semi-dependent (part of its essential needs intentionally provided by humans) and either semi-restricted or unrestricted; and (iv) feral dog: independent (none of its essential needs intentionally provided by humans) and unrestricted. This classification, however, is biased toward an anthropic perspective as it considers dogs' status only in relation to human dependency and control.

Boitani and Fabbri (1983) and Boitani *et al.* (1995) proposed a slightly different classification (four categories: owned restricted, owned unrestricted, stray and feral dogs) to account for the individual ontogeny of dog–human association, i.e. the nature and development of a social bond with humans, and the ecology of dogs at varying degrees of human dependency. Under this classification, the first two categories are similar to WHO's (1988) categories, i.e. restricted and family dogs. The family dogs have an owner on whom they depend, but may be left free to roam (Hsu *et al.*, 2003). The third category, stray dogs, includes dogs living in a human-dominated context: it is an heterogeneous group of dogs, comprising: (i) dogs that still have a social bond with humans, possibly abandoned or born into human settings; and (ii) dogs with different degrees of fear/tolerance towards humans, who are attracted near human settings by the availability of food and shelter resources regardless of whether these resources are intentionally provided by humans or are casually associated with them (e.g. handouts, refuse tips or garbage dumps for food, infrastructures for shelter, etc.). These dogs have also been called 'village dogs' (Macdonald and Carr, 1995; Ortolani and Coppinger, 2005) and have been consistently reported by many authors; those roaming many of the large cities of the Mediterranean basin (Istanbul, Alexandria) were described in the 18th century almost as a separate subspecies (cf. Brehm, 1893). The fourth category, feral dogs, includes all dogs living in a wild and free state with no direct food or shelter intentionally supplied by humans (Causey and Cude, 1980), showing no evidence of socialization to humans (Daniels and Bekoff, 1989a), but rather displaying a strong continuous avoidance of direct human contacts (Boitani and Ciucci, 1995; Boitani *et al.*, 1995) and a lifestyle restricted mainly to natural environments. Taking into account the social bond with humans and the ecological context adds important insights into free-ranging dogs' behaviour and has substantial implications for their classification and management.

Dogs' categories, however, are not homogeneous and the distinction among feral, stray/neighbourhood/village, and other free-ranging dogs is often a matter of degree (Nesbitt, 1975; Boitani and Ciucci, 1995), based on in-depth knowledge of their behaviour and ecology. Moreover, dogs from different categories can form temporary heterogeneous groups that are difficult to classify, and individual animals can change their status during their life depending on local ecological and social contexts, as well as following management interventions. Regardless of the names given to the various dog categories, most authors agree that 'owned', 'stray' and 'feral' dogs are not closed categories and that dogs may change their status throughout their life (Scott and Causey, 1973; Nesbitt, 1975; Hirata *et al.*, 1987; Daniels, 1988; Daniels and Bekoff, 1989a; Boitani *et al.*, 1995). The change that appears to be most difficult to recognize in the field is between the stray/neighbourhood/village dogs and the feral dogs, especially as it requires conceptual clarity on the dynamics of the feralization process. Feralization can be viewed from two different perspectives: (i) at the population level, as the domestication process in reverse (Hale, 1969; Brisbin, 1974; Price, 1984); and (ii) at individual level, as a behavioural process (Daniels and Bekoff, 1989c), as shown by Boitani *et al.*'s (1995) observation of feralization occurring within an individual's lifetime. The two perspectives are not mutually exclusive and imply different temporal scales as well as different theoretical and research approaches (Daniels and Bekoff, 1989c; Boitani and Ciucci, 1995).

A dog can change status as the result of several causes acting in complex interactions (Fig. 9.1). An owned dog can become stray by increasingly loosening restriction, by abandonment or simply by being born to a stray mother (Beck, 1975; Perry, 1993; Hsu *et al.*, 2003). A stray dog can become feral when forced out of a human environment or when accepted by a feral group (Daniels, 1988; Daniels and Bekoff, 1989a, c; Boitani *et al.*, 1995), like the majority of group members studied by Boitani *et al.* (1995). In changing status, dogs pass through intermediate behaviours depending on local social and ecological conditions, and the process may take a significant portion of an individual's life-span or may never be completed. This may be the case of many stray/neighbourhood/village dogs that live in the marginal areas of large human settlements, especially in tropical regions (Pal *et al.*, 1998a, b, 1999; Ortolani and Coppinger, 2005).

Daniels and Bekoff (1989c) suggested that feralization occurs through the development of a fear response to humans and they implied that the reverse process is possible. Indeed, dogs that shifted to a feral status can return to condi-

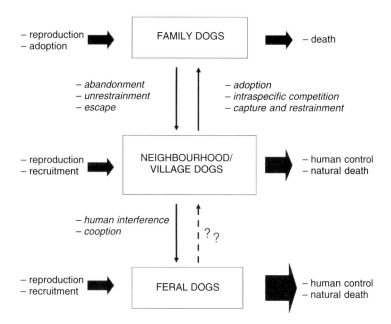

Fig. 9.1. A model of the feralization process. Population input and output mechanisms for each dog category (horizontal arrows) and the possible factors facilitating shifting status between different dog categories (vertical arrows).

tions of acceptance or tolerance of humans, as observed by Boitani *et al.* (1995) and experimentally shown by one of us by rehabilitating a feral dog to a domestic status (P. Ciucci, unpublished data). However, dogs that have not experienced contacts with humans during the first 2–3 weeks of life, will remain human-shy and may never be able to develop a full social bond with humans despite their socio-cognitive predisposition to attend to human signals (Hare *et al.*, 2002; see also Chapter 4 in this volume). The contrary, changing from stray to owned status, is more than often demonstrated by the many stray dogs adopted by humans.

The Behaviour of Neighbourhood/Village Dogs and their Interactions with People

Free-ranging dogs that associate with human settlements (hence village dogs) all share similar behavioural-ecological characteristics. They live mostly unrestrained, roving through the streets of villages, towns or cities, feeding mainly on human refuse and garbage, and using buildings, porches, vehicles or other human-built structures for shelter. Researchers who studied them in different parts of the world found that many village dogs actually have owners (Beck, 1975; Daniels, 1983a; Boitani and Racana, 1984; Brooks, 1990; DeBalogh *et al.*, 1993; Butler and Bingham, 2000; Ortolani and Coppinger, 2005), although they are not family pets, as we might think of by a Western concept. These dogs tend to associate with particular households (Boitani and Racana, 1984; DeBalogh *et al.*, 1993), even when the home-owners claim that the dogs don't belong to them (Ortolani and Coppinger, 2005). Most of them are solitary, although pairs or trios are not unusual, while large groups are fairly uncommon (notable exceptions are temporary gatherings of males around an oestrous female (Pal *et al.*, 1999) or dogs gathering at a common food source (Macdonald and Carr, 1995)) (Beck, 1975; Berman and Dunbar, 1983; Daniels, 1983a; Boitani and Racana, 1984; Hirata *et al.*, 1986; Daniels and Bekoff, 1989b; Macdonald and Carr, 1995; Ortolani and Coppinger, 2005). They are mainly active in the early hours of the morning and in the evening (Beck, 1975; Berman and Dunbar, 1983; Daniels, 1983a; Boitani and Racana, 1984; Hirata *et al.*, 1986); periods that often coincide with cooler temperatures of the day, especially in tropical regions. The majority of them are likely to be males, as the adult sex ratio is skewed in their favour, often in a ratio of two males for each female, but sometimes males can be as much as five times more abundant than females (Beck, 1973; Daniels, 1983b; Boitani and Racana, 1984; WHO, 1988; Daniels and Bekoff, 1989b; Brooks, 1990; DeBalogh *et al.*, 1993; Macdonald and Carr, 1995; Pal *et al.*, 1998a, b; Butler and Bingham, 2000; Ortolani and Coppinger, 2005).

Village dogs may have been the earliest canines to associate with people in a permanent way. Some 15,000 years ago, when humans traded a nomadic lifestyle for a sedentary one and the first villages were built, human-generated refuse began piling up around them: leftovers of family meals, butchering scraps, carcasses, by-products of farm animals, as well as human excrement. This newly created resource, the garbage dump, opened up a niche appealing especially to scavenger-prone carnivore species, with one caveat: tolerance to human proximity would have been the price to pay for admission. As Coppinger and Coppinger (2001) have suggested, dog domestication might indeed have taken this route: individuals who displayed lower flight distances towards humans around garbage dumps would have been selected to exploit this resource, and dogs' long-lasting co-existence with people would have thus begun (see also Chapters 2 and 3 in this volume for a further, in-depth, discussion of the timing and process of dog domestication). There is plenty of evidence today that village dogs gain food benefits by associating with humans (Beck, 1973; Daniels, 1983a; Boitani and Racana, 1984; Pal *et al.*, 1998a, b; Brooks, 1990; Macdonald and Carr, 1995; Butler and Bingham, 2000). In fact, village dogs may depend entirely on human garbage for survival, as their hunting skills are not very effective (see Boitani and Ciucci, 1995). The reward for trading a hunting life for one of scraps may be that village dogs are rarely found in poor body condition (DeBalogh *et al.*, 1993; Ortolani and Coppinger, 2005).

Shelter is another benefit supplied by villages, either intentionally or unintentionally provided by people, which dogs require for protection against harsh climatic conditions, aggressive conspecifics or carnivore competitors, and/or for hiding young from danger. But sharing a niche with humans has its costs. One characteristic shared by village dog populations around the world is a high rate of infant mortality: as many as 75% of litters can be destroyed by people (Macdonald and Carr, 1995) and often less than 20% of pups reach adulthood (WHO, 1988; Pal *et al.*, 1998a). Selective culling of female pups in litters is another common practice (A. Ortolani, 2003, personal observation; Daniels and Bekoff, 1989b), which most likely causes the biased sex ratio in the adult population. Finally, human culling, disease and car accidents together can cause more than 80% of adult mortality in some areas (DeBalogh *et al.*, 1993). Although these figures seem high and might give the impression that the costs outweigh the benefits for village dogs, similar mortality rates have also been reported for feral dogs (Boitani and Ciucci, 1995). More importantly, recorded population densities of village dogs range from 68 dogs/km^2 to 3700 dogs/km^2 in different parts of the world (WHO, 1988; Brooks, 1990), suggesting that high mortality rates are not putting these canines at risk of extinction. Although the diversity of dog–human relationships may mask the adaptive value of several dogs' traits, it appears that, at the population level, the costs of living with humans are probably negligible.

The association of village dogs with human households, although driven by the presence of a food source, might predispose them to become adopted and eventually become 'family dogs'. However, it does not imply that village dogs seek companionship with people, as in the popular image of a stray looking for an owner, nor does it imply that all village dogs could become family pets. In a recent study of human–dog interactions in Ethiopia, Ortolani and Coppinger (2005) found that most dogs avoided humans by fleeing when approached and that only a small minority (4%) would approach an observer. Moreover, many dogs (22%)

displayed alert/alarm behavioural responses, such as barking and growling, upon seeing an unfamiliar person, while some (11%) were downright aggressive, attempting to attack the observer. Dogs were more likely to vocalize if they were found near a household, suggesting that they might have been defending a resource or territory. Boitani and Racana (1984) also reported that village dogs barked near houses and livestock. Most Ethiopian village dogs could only be approached at a distance of 5 m or more, after which the majority either fled or displayed aggressive behaviour (Ortolani and Coppinger, 2005). Interestingly, such dogs tended to be alone or in pairs, while dogs encountered in small groups displayed mainly neutral, or indifferent, behaviours towards the observer. All aggressive dogs vocalized towards the observer before reacting, suggesting that, even though they might just have been bluffing, it is a wise person who pays attention to dogs' warning signs!

The majority of village dogs studied in different surveys were usually encountered in the streets and their main activity was either lying down resting or travelling around the village (Beck, 1975; Berman and Dunbar, 1983; Ortolani and Coppinger, 2005). Intraspecific agonistic interactions were not common but they appeared to peak during the oestrous season and usually between males (Daniels, 1983b; Pal *et al.*, 1999). Studies of dog–human interactions are scarce, and they are likely to be influenced by the general experience that dogs have formed with people habitually encountered in their environment. People's attitudes towards dogs vary widely among cultures, individual predispositions and even religious beliefs (Wandeler *et al.*, 1993). Curiously, in Ethiopia, Ortolani and Coppinger (2005) found that in predominantly Muslim towns the majority of dogs tended to avoid or have aggressive reactions towards an unfamiliar observer, while in predominantly Christian Orthodox villages the majority of dogs showed neutral reactions towards the same observer. Obviously, there are many factors, other than religion, that could be involved in explaining differences in dogs' reactions to people in those villages, but these results do suggest that people's attitudes may influence village dog behaviour.

Clearly, more studies quantifying behavioural responses of village dogs towards people in different areas of the world are needed, as well as more refined measures to quantify people's attitudes towards dogs in different cultures. Such studies would be particularly important in areas where potentially fatal dog-transmitted diseases, such as rabies, are still prevalent (see WHO, 2001). Although data are limited, village dogs may be the most abundant category of dog in the world (Beck, 1975; WHO, 1988) and, so far, they have been the most overlooked of all canines. Yet they hold the key to understanding how a predator evolved into the most faithful human companion. Studying the behaviour of village dogs is of critical importance to shed light on the complexities of human–dog interactions in our society.

The Behaviour of Feral Dogs

The behaviour and ecology of feral dogs result from the complex interaction of a suite of biological traits which still resemble those of their ancestors (wolves), especially concerning their ecological flexibility and the great variety of artificial and natural environments that they can live in. Many of the ancestral traits, such as a group-living tendency, territoriality, predatory instincts and a large degree of ecological flexibility, are still evident in the dogs' biology, but most of these traits appear void of their original adaptive value and may represent 'evolutionary inertia' or artificial selection epiphenomena (Boitani and Ciucci, 1995). If the process of domestication had the effect of enhancing the behavioural and ecological flexibility that allowed dogs to survive in a wide range of semi-natural environments, it also reduced their overall fitness to cope with long-term wild habitat conditions (Price, 1984). Feral dogs are generally not reproductively self-sustaining: they suffer from high rates of juvenile mortality; they depend indirectly upon humans for food, co-opting new individuals, and space; and their demography appears dominated by stochastic and unpredictable mechanisms (Boitani et al., 1995). Most studies of feral dogs' behaviour and ecology have been carried out on individuals which had been feral only for a few generations, and thus these generalizations may not apply to feral dogs such as dingoes and pariah dogs, which have been living under the effects of natural selection forces for longer periods of time.

Social ecology

The social structure of feral dogs is an aggregation of monogamous breeding pairs and their associates (pups and/or subadults of pair members), which is substantially different from the highly structured hierarchy of wolf packs, where dominance is respected both in terms of privilege (e.g. 'pecking order') and initiative (e.g. travelling, hunting, territorial defence, reproduction, etc.) (Mech and Boitani, 2003). Wolves' pack size is regulated also through social control of reproducing individuals (Packard, 2003), a mechanism which is apparently absent in dogs (Boitani et al., 1995). Intraspecific agonistic behaviour of feral dogs seems to be restricted primarily at the individual level without extensive effects on the social organization of the group, such as forcing a hierarchy and building a robust cohesiveness among group members. Scott and Fuller (1965) suggested that the agonistic behaviour of dogs toward humans is strongly influenced by their early contacts with humans and this might be particularly relevant when interpreting the behaviour of feral dogs recruited from the house and stray categories. Therefore, observations of feral dog behaviour without knowledge of the complete individual history of all group members are not sufficient to understand the causes and mechanisms of intraspecific relationships.

Members of feral dogs' groups are generally not related (Scott and Causey, 1973; Nesbitt, 1975; Causey and Cude, 1980; Berman and Dunbar, 1983; Daniels and Bekoff, 1989a, b; Boitani et al., 1995). The basic social unit is the breeding pair but the social bond is only loosely extended to the rest of the group members without the complex rules that regulate pack life in most wild canids (Kleiman and Eisenberg, 1973; Bekoff et al., 1984; Gittleman, 1989). In particular, the differences with wolf packs, which are highly cohesive one-family units that hunt, rear young and protect a communal territory as a stable group (Mech and Boitani, 2003), are obvious and prompted Boitani and Ciucci (1995) to propose the term 'group', as more appropriate than pack, for a feral dogs' social unit. Most studies of feral dogs found that group size ranges from two to six animals (two to five in Alabama: Scott and Causey, 1973; two to six in Alabama: Causey and Cude, 1980; two to four in Arizona: Daniels and Bekoff, 1989b; five to six in Illinois: Nesbitt, 1975; three to six in Italy: Boitani et al., 1995), in contrast with the smaller social unit of neighbourhood/village dogs, which are reported mostly in pairs or alone (see above on village/neighbourhood dogs). Macdonald (1983) suggested that in carnivores the quantity and distribution of food resources are the main determinants of group size. In wolves, pack size appears to be affected primarily by prey abundance, as changes in prey availability correspond to proportional changes in pack size (Mech and Boitani, 2003). It could be speculated that neighbourhood/village dogs would have little advantage from living in groups because resources are more scarce and scattered than those available to feral dogs, but also because cooperative 'hunting' would not be an advantage when resources are plentiful and easy to obtain. Also, lack of natural predators or competitors in human settings would not offer particular advantages to larger groups of stray dogs (cf. Macdonald and Carr, 1995), a situation that could be different in the case of feral dogs. However, even if studies with accurate estimates of food resources were available, any attempt to draw theoretical generalizations would be little more than a speculation, as we believe that theoretical analyses of canid evolutionary strategies are of limited value when carried out on animals which have been living under both artificial and natural selection pressures.

Boitani et al. (1995) reported that in their study area garbage dumps supplied food in excess during all seasons, and group size appeared to be related more to social factors, such as group turnover, rather than ecological ones, even though communal resource defence against wolves and other feral dog groups could have been a factor. Although no evidence of any intrinsic regulatory mechanism was found, the size of the feral dog group studied in Italy between 1984 and 1987 remained fairly stable (Boitani et al., 1995), and all events affecting group size appeared mostly due to density-independent and external factors (i.e. casual human persecution, availability of stray dogs, climatic conditions, etc.) with the sole exception of the cooption of stray dogs into the group (see below). Mortality of most sexually mature individuals was caused by human interference, while newborns from feral parents contributed almost nil to long-term group stability. Group size was maintained stable only by recruiting new members from the village dog population; at the end of the study, all but one dog in the group were of stray origin. Recruitment of village dogs into the feral group occurred mostly (but not only) during the breeding period and in conjunction with the disruption of a breeding pair. A single adult, after loosing its mate, would actively coopt another sexually mature dog, and this new member would in turn become socially accepted by the entire feral group. Thus, under the conditions studied by

Boitani *et al.* (1995) the 'vacancy' in a pair bond seems to be the main mechanism triggering cooption, although more data are needed from different ecological contexts to allow for meaningful generalizations.

One related interesting question, which remains unanswered, concerns the mechanisms, if any, that regulate the upper limit of group size in feral dogs. It may be speculated that given unlimited resources the lack of a firm social hierarchy and strong social bonds poses no upper limit to the number of feral dogs that can associate in a group. However, if functionality (in hunting, territorial defence, offspring care, etc.) is a prerequisite for the existence of an effective unit, then an upper limit appears to be naturally imposed by the costs of enduring effective cooperation. Since dogs form less efficient functional units compared to wolves, this might in part explain the smaller size of feral dog groups (Scott and Causey, 1973; Nesbitt, 1975; Causey and Cude, 1980; Daniels and Bekoff, 1989b; Boitani and Ciucci, 1995; Boitani *et al.*, 1995). Moreover, the social structure of feral dogs may not allow for an efficient density-dependent mechanism of population regulation in relation to environmental and ecological conditions, making feral dog groups more susceptible to stochastic events and limiting factors. Even when predictable and abundant food resources are available to feral dogs, the higher number of females reproducing per group, with their di-oestrous cycle and overall negative energetic balance, the lack of non-reproducing 'auxiliaries', and the high pup/juvenile mortality all interact to determine low recruitment rates and to maintain group size through external recruitment. In feral dogs, therefore, group size and composition appear to be a function of food abundance and availability of potentially cooptable village dogs at the lower end and, at the other extreme, of individual dogs' physiology and social behaviour, which hardly translates into functional and cohesive larger groups.

Home-range and territoriality

Feral dogs spend most of their life within well-defined home-ranges whose internal portions (core areas) are often defended against intruders, although their territorial behaviour is highly variable depending on several environmental and human-related causes (Scott and Causey, 1973; Causey and Cude, 1980; Gipson, 1983; Daniels and Bekoff, 1989a; Boitani *et al.*, 1995; Meek, 1999). Home-range sizes obtained by radio-telemetry studies range from 4.4–10.4 km^2 for three groups in east-central Alabama (Scott and Causey, 1973), to 18.7 km^2 in Alabama (Causey and Cude, 1980), 57.8 km^2 in central Italy (Boitani *et al.*, 1995) and 70 km^2 in Alaska (Gipson, 1983). Home-range size does not appear to be closely related to group size, but rather is dependent on the spatial patterns of key resource sites, such as denning areas, refuge areas, garbage dumps and other food sources. Dogs may show seasonal variations in patterns of range utilization, using smaller portions at different times. Daniels and Bekoff (1989a) and Scott and Causey (1973) related these core-area variations to the presence of dependent pups and to different energetic requirements of the group. In contrast, Boitani *et al.* (1995) found that several other factors could possibly be involved in shifting of the core areas within the home-range: the exploitation of temporary food resources (i.e. a large livestock carrion), disturbance caused by humans, denning activities, previous spatial-use patterns of newly recruited dogs, unpredictable fluctuation in food availability at dumps, and possible interference by wolves. In the same study, it was suggested that shifting of core areas and seasonal ranges were not necessarily only a consequence of direct environmental changes, in fact previous knowledge of the area by a new member of the group could affect the entire group's spatial behaviour.

Home-range sizes tend to become smaller in neighbourhood/village dogs living closer to or within areas inhabited by humans (from 2–11 ha up to 61 ha) (Beck, 1973; Fox *et al.*, 1975; Berman and Dunbar, 1983; Daniels, 1983a; Santamaria *et al.*, 1990), confirming that predictability and quantity of food sources as well as small group sizes are among the determinants of ranging behaviour. In feral dogs, home-ranges, or parts of them, especially core areas, food sources and den sites, are actively defended using scent marking, vocalizations (barking) and aggressive behaviour. Both Macdonald and Carr (1995) and Boitani *et al.* (1995) reported for the same study area in central Italy that defensive behaviour was consistently shown within the entire core areas and during the whole year. The higher frequency of territorial behaviour found in Italy compared to other studies (Scott and Fuller, 1965; Bekoff, 1979; Berman and Dunbar, 1983; Boitani and Racana, 1984; Daniels and Bekoff, 1989a) might be related to a higher level of integration within the group, a higher degree of isolation from other dogs, and to food resources being concentrated mostly in localized patches at the dumps. In addition, an unknown effect might be due to the typical range defending behaviour of the Maremma dog, a large guarding dog breed which was the dominant type of that feral group (Boitani *et al.*, 1995). Interspecific territorial behaviour among wild canids is known to occur (e.g. coyote–wolf, fox–coyote) and it should be expected in dogs and wolves that share the same ranges. In central Italy, dogs and wolves share an almost identical niche and compete for the same food resources (Boitani, 1983), but their territories only partially overlap (Boitani *et al.*, 1995). In Abruzzo, the territorial core-areas of feral dogs were closer to human settlements where wolf presence was lowest, and they were located in the interstice between two neighbouring wolf territories (Boitani *et al.*, 1995). These observations suggest that wolves in that particular area might have been an important component in shaping feral dogs' territory and in determining its location in relation to human settings.

In contrast to house and village dogs, feral dogs are most definitively active during nocturnal and crepuscular periods with a clear tendency for a bimodal activity pattern, similar to other wild canids (Perry and Giles, 1971; Scott and Causey, 1973; Causey and Cude, 1980; Boitani and Racana, 1984; Daniels and Bekoff, 1989a; Boitani *et al.*, 1995). Nevertheless, Nesbitt (1975) and Boitani *et al.* (1995) found that dogs could also travel during daytime, when human presence and interference is low, suggesting that feral dogs adopt nocturnal habits essentially to avoid, or minimize, human contacts. However, a great variety of wild carnivores show a bimodal pattern of activity, and it has been hypothesized that this innate behavioural trait is independent from environmental pressures and has not been altered to a great extent by artificial selection on dogs.

Food sources, hunting and predation

Given their social structure and their adaptability to a variety of habitats, it seems obvious to expect that feral dogs would display a diversified feeding ecology and include a wide range of food items in their diet. While in wolves the social structure of packs integrates all members in an efficient hunting unit, in feral dog groups leadership is more questionable and social bonds among individuals are more flexible, which might contribute to the inefficiency of feral dogs as predators compared to their wild canine counterparts (Scott and Causey, 1973; Nesbitt, 1975; Causey and Cude, 1980; Daniels and Bekoff, 1989b; Boitani et al., 1995; Macdonald and Carr, 1995; Butler et al., 2004). A variety of other factors can influence their feeding behaviour, such as group size and breed types, relative abundance and accessibility of wildlife and livestock, the level of human control and the availability of garbage dumps and alternate food sources easy to exploit. Group members' cultural traditions might also exert great influence on the group's feeding behaviour, determining hunting attitudes and ability as well as prey-type preference.

Feral dogs have long been accused of preying on wildlife and livestock and causing serious damage in a variety of geographic areas (see review in Boitani and Ciucci, 1995), but the supporting evidence is surprisingly scarce. Several studies (e.g. Scott and Causey, 1973; Nesbitt, 1975; Boitani et al., 1995) failed to document livestock depredation in spite of free-ranging livestock available to feral dogs in most of the study areas. Instead, as documented by Boitani et al. (1995) and Nesbitt (1975), livestock depredations were caused by free-ranging neighbourhood/village/stray dogs. Further evidence has been obtained to support feral dogs' predation on wildlife (i.e. deer, wild boar, hare, rabbit, etc.), although all studies failed to document any serious impact on wildlife populations (Perry and Giles, 1971; Scott and Causey, 1973; Gipson and Sealander, 1977; Causey and Cude, 1980; Federoff et al., 1994; Boitani et al., 1995; Herranz et al., 2000; Rouys and Theuerkauf, 2003; Butler et al., 2004). In Zimbabwe, Butler and du Toit (2002) found that dogs were primarily scavengers of human waste and animal carcasses, roaming into nature reserves and outcompeting other wild scavengers. Although predation on deer and boar populations might be of little management significance, feral dogs have been reported to prey on many endangered and rare species, causing serious conservation concerns: marine iguanas in the Galápagos Islands (Kruuk and Snell, 1981; Barnett and Rudd, 1983); capybaras in Venezuela's llanos (Macdonald, 1981); Indian porcupines in India (Chhangani, 2003); leatherback turtles on the Andaman and Nicobar Islands in the Indian Ocean (Andrews and Shanker, 2003); mountain gazelles in central Arabia (Dunham, 2001); and sable antelopes in Africa (Dott, 1986). Finally, in one report

Kamler *et al.* (2003) observed three feral dogs successfully attacking and killing a coyote, perhaps during an interspecific territorial dispute.

On the basis of all reviewed studies it seems reasonable to conclude that feral dogs show a generally low predatory attitude, a low kill rate (efficiency of predation), have potential limits in prey size and are indirectly dependent on humans for food. Notwithstanding the inherent weaknesses of their predatory behaviour, feral dogs can be a serious nuisance to local farmers and a threat to protected wildlife despite efforts to control their numbers, especially in protected areas (Johnson, 2002).

Breeding, denning and parental care

Of all traits affected by artificial selection, reproduction has been strongly manipulated to increase reproductive potential and to shorten generation time in dogs (Boitani and Ciucci, 1995). In the feral group studied by Boitani *et al.* (1995), all females reproduced giving the group full potential for demographic increase. Domestic dogs usually breed twice a year with little or no seasonal patterns, but feral dogs tend to concentrate their oestrus cycles in the spring in northern temperate environments (Gipson, 1972; Daniels and Bekoff, 1989b), and in the autumn, or post-monsoon period, in India (Chawla and Reece, 2002; Pal, 2003). Boitani *et al.* (1995) found on average 7.3 months (range 6.5–10 months) between oestrus periods; 50% of births occurred during February–May, whereas the others were scattered during the rest of the year. The spring peak of births is common and has obvious adaptive value among wild canids and many other wildlife species in the northern hemisphere. In feral dogs it might just be the remnant of an ancestral endogenous reproductive rhythm without any current adaptive value. More interesting is the lack of synchronization of the breeding females in the feral group found by Boitani *et al.* (1995), and the casual distribution throughout the year of the other 'non-spring' oestrus cycles, which might be resulting from artificial selection and disconnecting reproductive rhythms from natural photoperiod synchronization and social control (but see Macdonald and Carr, 1995).

In the same study, there was no indication of communal care of litters by the group and all females reared their pups alone, although they were often visited by other group members. Females located their dens near the group's traditional core areas and spent most of their time at the den but frequently visited the closest feeding sources. Even though food was abundant throughout the denning period, the lack of group members' support, in terms of vigilance and protection of the pups, during the mothers' frequent absences might have contributed to the high rate of infant mortality due to predation (Boitani *et al.*, 1995). Daniels (1988) and Daniels and Bekoff (1989b) reported that breeding females split from the group, denned and reared pups in isolation, though in the vicinity of the group. Alloparental care seems to be an adaptive behaviour in many social wild canids because it relieves the female from the burden of caring for her pups alone and it may increase protection from intruders and predators. Dogs appear to be the only canids without any form of paternal care (Macdonald and Carr, 1995), although Malm (1995) found that a substantial proportion of family-owned male dogs participate in caring for the pups, mainly by providing regurgitated food. Domestication and human assistance to reproduction might have played a significant role in altering parental behaviours as well as eliminating most of the social control on reproduction within the group.

In feral dogs, litter sizes range from 3.6 pups/litter ($n = 11$, Boitani *et al.*, 1995) to 5.5 pups/litter ($n = 17$, Macdonald and Carr, 1995), but pup survival rates are very low. Boitani *et al.* (1995) found that out of 40 pups, 28 (70%) died within 70 days of birth, nine (22.5%) died within 120 days, one (2.5%) within 1 year, and only two (5%) survived the age of 1 year. Similar results have been obtained from other studies (Scott and Causey, 1973; Nesbitt, 1975; Daniels and Bekoff, 1989b; Macdonald and Carr, 1995). Most mortality seems to occur during the period of early independence and may be due to: (i) the absence of communal helping and increased risks of predation when pups are left unattended, and when they begin to explore the areas surrounding the den site; (ii) a lowered maternal interest in offspring as the mother enters a new oestrus cycle; or (iii) hostile environmental conditions for litters born in periods other than spring or early summer. In short, the reproductive traits selected in domestic dogs are of low adaptive value back in the wild, where feral dogs suffer from a very inefficient reproductive mechanism, which tries to maximize production while minimizing newborn and juvenile survival (often less than 5% surviving to 1 year of age) to the point that feral populations appear unable to sustain themselves. In central Italy, the feral dogs in Boitani *et al.*'s (1995) study could not have maintained their numbers without continuously recruiting new group members from neighbourhood/village/stray dogs' populations.

Finally, a curious and so far unexplained aspect of feral dogs' demography is the often reported highly skewed sex ratio in favour of males. While in village dogs there might be an important effect of selective removal of females, in feral dogs it is difficult to expect a differential mortality rate for the two sexes outside artificial human interference. Boitani *et al.* (1995) found the overall litter composition highly skewed in favour of males (3.2:1) compared to the female-biased adult sex ratio, but they could not provide any conclusive explanation for the possible reasons of higher female survival rates.

Feral Dogs and Humans: a Less Obvious Form of Dependency

In controlled settings (i.e. family dogs), domestic dogs depend entirely on humans for all their biological, social and ecological requirements (food, health, range, housing, protection from predators, social relationships, reproduction, welfare, etc.). However, under some circumstances, this bond may become weaker and dogs can adjust to a looser relationship with humans either intentionally (by escaping and affiliating with other free-ranging dogs), accidentally (loss of owner, birth from a stray mother) or forcefully (abandonment). When this occurs (by human intervention), it is up to the dog obviously to decide if a looser lifestyle is more interesting, but this is not a field where scientists easily venture (but see Hart, 1995). What we can do instead is try to understand the causes, conditions and implications of dogs living partially or totally unrestricted.

Of the few generalizations we can draw from the studies discussed above, two offer relevant behavioural insights and point to the same management implication: first, an unrestricted lifestyle has its costs, and the looser the dog the higher the costs. Unpredictable food resources, occasional shelter, unsafe habitat, exposure to diseases and predators all contribute to determine high mortality rates in neighbourhood, village, stray and feral dogs to the point that, particularly in Western societies and temperate climates, these dog populations are often not reproductively self-sustaining. Second, even though these dogs are increasingly independent from humans for their immediate decisions (use of space and resources, activity, social relationships), they still depend heavily, though indirectly, on human activities for most of their key resources. This dependency may be more evident in neighbourhood and village dogs and less obvious in more elusive forms of free-ranging dogs. By the simple fact that stray-feral and true feral dogs tend to live in increasingly wild conditions avoiding contacts with humans, one could be tempted to equate them to other wild canids in terms of their ecological relationship with the environment, implying that they are regulated by natural ecological processes in the absence of human interference and representing ecological and evolutionary adaptive units. However, with possibly a few exceptions (e.g. wild dingoes: Price, 1984), which result from ecological and anthropological dynamics of thousands of years ago, no longer plausible in today's world, no true feral dog population appears to be ecologically and reproductively self-sustaining or, more importantly, to be reproductively isolated from other domestic dog categories. On the contrary, most feral dog populations represent ephemeral entities whose existence is linked to human activities to various degrees.

By reviewing studies on feral dogs carried out in different areas of the world (see above), several factors appear critical for their persistence. They underline the dependency of feral dogs upon humans and involve at least four aspects of their biology: nutrition, ecology, demography and management. First, the availability of abundant and predictable food sources appears a critical prerequisite for the establishment of feral dogs' populations and their persistence through time. The behavioural ecology of the studied feral dog groups is explained largely by the nature, dispersal and abundance of human-derived food resources (Scott and Causey, 1973; Nesbitt, 1975; Causey and Cude, 1980; Daniels, 1988; Daniels and Bekoff, 1989b; Boitani et al., 1995; Macdonald and Carr, 1995). Feral dogs, as other wild carnivores, are limited by bio-energetic constraints and, differently from neighbourhood and village dogs, they need to forage and hunt by themselves (i.e. no food is intentionally provided by humans). However, feral dogs do not appear to be effective predators and, in most situations, they thrive only where abundant and highly predictable food sources (i.e. garbage dumps) are available. In the area studied by Boitani *et al.* (1995), 3 years after the closure of one of the main garbage dumps no feral dogs could be found in their traditional core areas, although other factors could also have been potentially involved (illegal control, poisoning) (P. Ciucci, 2004, personal observation). As Macdonald and Carr (1995) pointed out, the large availability of food at garbage dumps, compared to scraps, handouts and other less abundant food sources in the villages, can be a factor involved in observing larger group sizes of feral dogs, suggesting that their social ecology as well is dictated by food sources of anthropogenic nature.

A second factor critical for the establishment of feral dogs is the availability of suitable habitats and space. Suitable for feral dogs generally means a natural environment void of human settlements and rich in retreat and denning areas but, simultaneously, where food sources are adequately dispersed and accessible (see above), and where human density and/or past extermination efforts correspond to low density or absence of wild predators (e.g. wolves). Third, since feral dog populations are often not reproductively self-sustaining and suffer high mortality, their persistence and demography are dependent upon the continuous availability of individuals to be recruited from other free-ranging dog categories (see Fig. 9.1), whose maintenance and dynamics is, in turn, linked to human activities. Finally, assuming all above conditions, feral dogs' survival is ultimately allowed where environmental and wildlife management regulations, as well as human health prescriptions, are either neglected or difficult to implement. Feral dogs have been found in areas with little tradition or enforcement of wildlife management (e.g. Daniels, 1988; Boitani *et al.*, 1995) or in remote and inaccessible sites (e.g. Nesbitt, 1975; Causey and Cude, 1980). To ensure conservation and functionality of natural communities and ecosystems, wildlife management protocols prescribe that dogs which escaped human restraint are controlled, similarly to exotic and alien species. Boitani *et al.* (1995) outlined guidelines for cost-effective, humane and non-lethal control of feral dogs, but the degree to which we want to adhere to, or are able to implement, these or other directions is a phenotypic expression of our society upon which feral dogs are again dependent. From this perspective, humans being part of feral dogs' ecology, we are responsible for their persistence as well as for maintaining the ecological conditions that allow for their survival.

Although they are more elusive and apparently independent from humans, feral dogs are susceptible to disappear where one or more of the above conditions are not met, at least with higher probabilities than other free-ranging dog categories which have closer behavioural and ecological bonds with humans. In conclusion, we would generally predict that the survival of feral dogs is more of a temporary and context-related phenomenon than previously thought, and that humans are more responsible for the dogs' lifestyle choices than the dogs themselves.

References

Andrews, H. and Shanker, K. (2003) A significant population of leatherback turtles in the Indian Ocean. *Kachhapa Newsletter* 6.

Barnett, B.D. and Rudd, R.L. (1983) Feral dogs of the Galapagos Islands: impact and control. *International Journal of Studies on Animal Problems* 4, 44–58.

Beck, A.M. (1973) *The Ecology of Stray Dogs: A Study of Free-Ranging Urban Animals.* York Press, Baltimore, Maryland.

Beck, A.M. (1975) The ecology of 'feral' and free-roving dogs in Baltimore. In: Fox, M.W. (ed.) *The Wild Canids.* Van Nostrand Reinhold, New York, pp. 380–390.

Bekoff, M. (1979) Scent-marking by free-ranging domestic dogs. *Biological Behaviour* 4, 123–139.

Bekoff, M., Daniels, T.J. and Gittleman, J.L. (1984) Life history patterns and the comparative social ecology of carnivores. *Annual Review of Ecology and Systematics* 15, 191–232.

Berman, M. and Dunbar, I. (1983) The social behaviour of free-ranging suburban dogs. *Applied Animal Ethology* 10, 5–17.

Boitani, L. (1983) Wolf and dog competition in Italy. *Acta Zoologica Fennica* 174, 259–264.

Boitani, L. and Ciucci, P. (1995) Comparative social ecology of feral dogs and wolves. *Ethology, Ecology and Evolution* 7, 49–72.

Boitani, L. and Fabbri, M.L. (1983) Censimento dei cani in Italia con particolare riguardo al fenomeno del randagismo. *Ricerche di Biologia della Selvaggina* (INFS, Bologna, Italy) 73, 1–51.

Boitani, L. and Racana, A. (1984) Indagine eco-etologica sulla popolazione di cani domestici e randagi di due comuni della Basilicata. *Silva Lucana* (Bari, Italy) 3, 1–86.

Boitani, L., Francisci, F., Ciucci, P. and Andreoli, G. (1995) Population biology and ecology of feral dogs in central Italy. In: Serpell, J. (ed.) *The Dog: Its Evolution, Behaviour and Interactions with People.* Cambridge University Press, Cambridge, pp. 217–244.

Brehm, A. (1893) *Tierleben.* Liepzig-Wien, 4 vols.

Brisbin, I.L. Jr. (1974) The ecology of animal domestication: its relevance to man's environmental crises – past, present and future. *Association of Southeastern Biologists Bulletin* 21, 3–8.

Brooks, R. (1990) Survey of the dog population of Zimbabwe and its level of rabies vaccination. *Veterinary Record* 127, 592–596.

Butler, J.R.A. and Bingham, J. (2000) Demography and dog–human relationships of the dog population in Zimbabwean communal lands. *Veterinary Record* 147, 442–446.

Butler, J.R.A. and du Toit, J.T. (2002) Diet of free-ranging domestic dogs (*Canis familiaris*) in rural Zimbabwe: implications for wild scavengers on the periphery of wildlife reserves. *Animal Conservation* 5, 29–37.

Butler, J.R.A., du Toit, J.T. and Bingham, J. (2004) Free-ranging domestic dogs (*Canis familiaris*) as predators and prey in rural Zimbabwe: threats of competition and disease to large wild carnivores. *Biological Conservation* 115, 369–378.

Causey, M.K. and Cude, C.A. (1980) Feral dog and white-tailed deer interactions in Alabama. *Journal of Wildlife Management* 44, 481–484.

Chawla, S.K. and Reece, J.F. (2002) Timing of oestrus and reproductive behaviour in Indian street dogs. *Veterinary Record* 150, 450–451.

Chhangani, A.K. (2003) Dogs (*Canis familiaris*) hunting in the wild at Jodhpur, Rajasthan. *Journal of the Bombay Natural History Society* 100, 617.

Coppinger, R. and Coppinger, L. (2001) *Dogs: A Startling New Understanding of Canine Origin, Behavior and Evolution.* University of Chicago Press, Chicago, Illinois.

Daniels, T.J. (1983a) The social organization of free-ranging urban dogs: I. Non-estrous social behaviour. *Applied Animal Ethology* 10, 341–363.

Daniels, T.J. (1983b) The social organization of free-ranging urban dogs: II. Estrous groups and the mating system. *Applied Animal Ethology* 10, 365–373.

Daniels, T.J. (1988) Down in the dumps. *Natural History* 97, 8–12.

Daniels, T.J. and Bekoff, M. (1989a) Spatial and temporal resource use by feral and abandoned dogs. *Ethology* 81, 300–312.

Daniels, T.J. and Bekoff, M. (1989b) Population and social biology of free-ranging dogs, *Canis familiaris*. *Journal of Mammalogy* 70, 754–762.

Daniels, T.J. and Bekoff, M. (1989c) Feralization: the making of wild domestic animals. *Behavioural Processes* 19, 79–94.

DeBalogh, K.K.I.M., Wandeler, A.I. and Meslin, F.-X. (1993) A dog ecology study in an urban and semi-rural area of Zambia. *Onderstepoort Journal of Veterinary Research* 60, 437–443.

Dott, A. (1986) Sable antelope falls victim to domestic dogs. *African Wildlife* 40, 138.

Dunham, K.M. (2001) Status of a reintroduced population of mountain gazelles *Gazella gazella* in central Arabia: management lessons from an arid land reintroduction. *Oryx* 35, 111–118.

Federoff, N.E., Jakob, W.J. and Bauer, W.C. (1994) Female feral dog and two pups kill deer fawn at the Patuxent Wildlife Research Center, Laurel, Maryland. *Maryland Naturalist* 38, 1–2.

Fox, M.W., Beck, A.M. and Blackman, E. (1975) Behaviour and ecology of a small group of urban dogs (*Canis familiaris*). *Applied Animal Ethology* 1, 119–137.

Gipson, P.S. (1972) The taxonomy, reproductive biology, food habits, and range of wild *Canis* (Canidae) in Arkansas. PhD Dissertation, University of Arkansas, Fayetteville.

Gipson, P.S. (1983) Evaluation and control implications of behaviour of feral dogs in Interior Alaska. In: Kaukeinen, D.E. (ed.) *Vertebrate Pest Control and Management Materials: 4th Symposium.* ASTM Special Technical Publication, Philadelphia, pp. 78–92.

Gipson, P.S. and Sealander, J.A. (1977) Ecological relationships of white-tailed deer and dogs in Arkansas. In: Philips, R.L. and Jonkel, C. (eds) *Proceedings of 1975 Predator Symposium. Bulletin of Montana Forest Conservation Experimental Station.* University of Montana, Missoula, Montana, pp. 3–16.

Gittleman, J.L. (1989) *Carnivore Behaviour, Ecology and Evolution.* Chapman and Hall, London.

Hale, E.B. (1969) Domestication and the evolution of behaviour. In: Hafez, E.S.E. (ed.) *The Behaviour of Domestic Animals.* Bailliere Kendall, London, pp. 22–42.

Hare, B., Brown, M., Williamson, C. and Tomasello, M. (2002) The domestication of social cognition in dogs. *Science* 298, 1634–1636.

Hart, L.A. (1995) Dogs as human companions: a review of the relationship. In: Serpell, J. (ed.) *The Dog: Its Evolution, Behaviour and Interactions with People.* Cambridge University Press, Cambridge, pp. 161–178.

Herranz, J., Yanes, M. and Suarez, F. (2000) Relationships among the abundance of small game species, their predators, and habitat structure on Castilla-La Mancha (Spain). *Ecologia* 14, 219–233.

Hirata, H., Okuzaki, M. and Obara, H. (1986) Characteristics of urban dogs and cats. In: Obara, H. (ed.) *Integrated Studies in Urban Ecosystems as the Basis of Urban Planning. I. Special Research Project on Environmental Science* (B276-R15-3). Ministry of Education, Tokyo, pp. 163–175.

Hirata, H., Okuzaki, M. and Obara, H. (1987) Relationships between men and dogs in urban ecosystem. In: Obara, H. (ed.) *Integrated Studies in Urban Ecosystems as the Basis of Urban Planning. II. Special Research Project on Environmental Science* (B334-R15-3). Ministry of Education, Tokyo, pp. 113–120.

Hsu, Y., Severinghaus, L.L. and Serpell, J.A. (2003) Dog keeping in Taiwan: its contribution to the problem of free-roaming dogs. *Journal of Applied Animal Welfare Science* 6, 1–23.

Johnson, M.R. (2002) A new capture pen for Caribbean feral dog packs. *Intermountain Journal of Sciences* 8, 255.

Kamler, J.F., Keeler, K., Wiens, G., Richardson, C. and Gipson, P.S. (2003) Feral dogs, *Canis familiaris*, kill coyote, *Canis latrans*. *Canadian Field-Naturalist* 117, 123–124.

Kleiman, D.G. and Eisenberg, J.F. (1973) Comparisons of canid and felid social systems from an evolutionary perspective. *Animal Behaviour* 21, 637–659.

Kruuk, H. and Snell, H. (1981) Prey selection by feral dogs from a population of marine iguanas (*Amblyrhynchus cristatus*). *Journal of Applied Ecology* 18, 197–204.

Macdonald, D.W. (1981) Dwindling resources and the social behaviour of Capybaras (*Hydrochoerus hydrochaeris*). *Journal of Zoology* 194, 371–391.

Macdonald, D.W. (1983) The ecology of carnivore social behaviour. *Nature* 301, 379–384.

Macdonald, D.W. and Carr, G. (1995) Variation in dog society: between resource dispersion and social flux. In: Serpell, J.A. (ed.) *The Dog: Its Evolution, Behaviour and Interactions with People.* Cambridge University Press, Cambridge, pp.199–216.

Malm, K. (1995) Regurgitation in relation to weaning in the domestic dog: a questionnaire study. *Applied Animal Behaviour Science* 43, 111–122.

Mech, L.D. and Boitani, L. (2003) Social ecology of the wolf. In: Mech, L.D. and Boitani, L. (eds) *Wolves: Behavior, Ecology and Conservation.* University of Chicago Press, Chicago, Illinois, pp. 1–34.

Meek, P.D. (1999) The movement, roaming behaviour and home range of free-roaming domestic dogs, *Canis lupus familiaris*, in coastal New South Wales. *Wildlife Research* 26, 847–855.

Nesbitt, W.H. (1975) Ecology of a feral dog pack on a wildlife refuge. In: Fox, M.W. (ed.) *The Wild Canids.* Van Nostrand Reinhold, New York, pp. 391–395.

Ortolani, A. and Coppinger, R. (2005) Behavioral ecology and human interactions of village dogs: a survey in Ethiopia and a comparative review. *Applied Animal Behaviour Science* (under review).

Packard, J.M. (2003) Wolf behaviour: reproductive, social, and intelligent. In: Mech,

L.D. and Boitani, L. (eds) *Wolves: Behavior, Ecology and Conservation.* University of Chicago Press, Chicago, Illinois, pp. 35–65.
Pal, S.K. (2003) Reproductive behaviour of free-ranging rural dogs in West Bengal, India. *Acta Theriologica* 48, 271–281.
Pal, S.K., Gosh, B. and Roy, S. (1998a) Dispersal behaviour of free-ranging dogs (*Canis familiaris*) in relation to age, sex, season and dispersal distance. *Applied Animal Behaviour Science* 61, 123–132.
Pal, S.K., Gosh, B. and Roy, S. (1998b) Agonistic behaviour of free-ranging dogs (*Canis familiaris*) in relation to season, sex and age. *Applied Animal Behaviour Science* 59, 331–348.
Pal, S.K., Gosh, B. and Roy, S. (1999) Inter- and intra-sexual behaviour of free-ranging dogs (*Canis familiaris*). *Applied Animal Behaviour Science* 62, 267–278.
Perry, B.D. (1993) Dog ecology in eastern and southern Africa: implications for rabies control. *Onderstepoort Journal of Veterinary Research* 60, 429–436.
Perry, M.C. and Giles, R.H. (1971) Free running dogs. *Virginia Wildlife* 32, 17–19.
Price, E.O. (1984) Behavioural aspects of animal domestication. *The Quarterly Review of Biology* 59, 1–32.
Rouys, S. and Theuerkauf, J. (2003) Factors determining the distribution of introduced mammals in nature reserves of the southern province, New Caledonia. *Wildlife Research* 30, 187–191.
Santamaria, A., Passannanti, S. and Di Franza, D. (1990) Censimento dei cani randagi in un quartiere di Napoli. *Acta Medica Veterinaria* 36, 201–213.
Scott, J.P. and Fuller, J.L. (1965) *Genetics and the Social Behavior of the Dog.* University of Chicago Press, Chicago, Illinois.
Scott, M.D. and Causey, K. (1973) Ecology of feral dogs in Alabama. *Journal of Wildlife Management* 37, 253–265.
Wandeler, A.I., Matter, H.C., Kappeler, A. and Budde, A. (1993) The ecology of dogs and canine rabies: a selective review. *Revue Scientifique et Technique de Office Internationale des Epizooties* 12, 51–71.
World Health Organization (1988) Report of WHO consultation on dog ecology studies related to rabies control. Ref. WHO/Rab.Res/88.25, Geneva.
World Health Organization (2001) Rabies: Fact Sheet N.99, revised June 2001. http://www.WHO.int/rabies/rabies_fs99

10 Evolutionary Aspects on Breeding of Working Dogs

Rolf Beilharz

Introduction

Consider the animals on any undisturbed location on Earth. There are many species and each species seems to be well matched to its particular niche in the environment. Ecologically, all animal species have relationships with many others and with plants, fungi and other microorganisms. Plants derive their energy from the Sun. Animals have to obtain their energy from other forms of life, either while it is still alive (e.g. living plants or prey animals) or from decaying carcasses and even dung. Everyone agrees that life depends on energy and that the energy required by animals for life must derive from other life in the individual animal's environment.

Energy is not the only environmental resource required for animal life. Many other environmental aspects must also match appropriately (e.g. nesting places, suitable climate and so on). However, energy for the metabolism necessary for all aspects of life is by far the most important resource necessary for animal life. Why is energy, or why are environmental resources generally, so well matched to animal life? Simply because energy and all environmental resources in the environment exert 'natural selection' on populations of animals.

Charles Darwin (1859) introduced the concept of natural selection. He learned from animal breeders that selecting and breeding from desired animals leads to better livestock. This reassured him that natural selection by the environment was having similar effects on all forms of life. He saw that different environments select different varieties in organisms. As more young are produced than are needed to replace deaths in a species, not all young will survive. Those that utilize the available resources better than others have a greater probability of survival. And this happens in every generation. Thus, if there is a mechanism of inheritance which allows the better characteristics to be passed on to progeny, those individuals better matched (or better adapted) to the environment will increase in numbers and eventually displace all others.

Natural selection is going on all the time. As different geographical locations exert different selection pressures on each local population, the local populations will adapt each in its own way and this may go so far that interbreeding between these different populations stops. In this way populations in different geographical areas can become different species. With the world being very much older than had been calculated from the Bible, this continual natural selection was enough to have achieved the variety of species making up the different forms of life now on Earth. The concept of natural selection nowadays makes sense to most people.

Darwin argued in terms of whole organisms, or specific features (characters or traits) of whole organisms, which match the environment. The whole organism and its parts are called the phenotype. We must never forget that natural selection acts on phenotypes. While Darwin did not know the mechanism of inheritance, he assumed that there must be some such mechanism present.

Mendel (read 1855, printed 1866) discovered the mechanism of inheritance. The importance of his work was recognized only in 1900. He realized that there were factors, now called genes, which move from parents to progeny in predictable ways (see Chapter 5 in this volume for an in-depth discussion of genetics). The rediscovery of Mendel's genes led to an ever increasing understanding of what happens inside the nuclei of cells. Chromosomes, double stranded strings, were identified as the carriers of genes. Genes were mapped on to the chromosomes using crossover experiments. If a gene occurs in more than one version, with different effects, the different versions are called alleles. From early in the 20th century, the theory associated with Mendel's genes was expanded to include the situation where organisms and their genes occur in populations. This is population genetics. It was followed rapidly by quantitative genetics, which describes gene action in those characters where many genes with small effects are responsible for the phenotype. In both cases the theory describes the situation in terms of genes and postulates what must happen to the genes involved and their alleles, in populations and in traits where there is no single major gene acting, respectively. It is clearly understood that the genes produce, or at least influence, the phenotype. Quantitative genetics recognizes that the environment also affects the resulting phenotype, although the environment described is restricted to those environments that are deviations from the population mean.

In 1953, the chemical structure of the material of the chromosomes (DNA) was elucidated. Since then, biochemistry at the molecular level has progressed rapidly. This 'biotechnology' has allowed individual genes and whole genomes to be defined. A genome is the molecular description of the whole genetic material in the chromosomes of an individual and hence of its species. In other words, we can now describe the genetic information about individuals or species precisely in terms of biochemical molecules. We can also manipulate this material biochemically, and transfer genetic instructions from one species to another (even from bacteria to humans). Scientifically this is a huge success and many people see this development as entirely positive. Biotechnology has, however, pushed the importance of natural selection aside. The earlier clear understanding that the environment selected whole organisms and adapted them to their specific environments is largely ignored in the modern paradigm of biology.

Just as natural selection adapts all species to their environments, humans have also selected animals and plants for their own purposes. The initial domestication of plants and animals was for a long time a process in which human activities simply added their effect to that of natural selection. Merely by building dwellings, humans provided environmental habitats in which certain species thrived (e.g. pigeons, sparrows, mice and rats). Food wasted by humans also increased populations of wolves and other animals and birds in their vicinity. Wolves later became domesticated as dogs (see Chapters 2 and 3 in this volume). Dogs and humans combined to hunt game animals, and merely by favouring the progeny of the best hunting dogs, humans caused breeds of hunting dogs to evolve. This process of initial domestication was an evolutionary process in which natural selection on fitness of animals continued but was biased by human activities, largely by the environmental changes created by humans. It is only in the last 200 years that science has begun to alter domestic plants and animals systematically, without, and increasingly in opposition to, the positive action of natural selection.

In plants, biotechnology has facilitated direct manipulation of the genetic material (DNA) to the point where genetic material can be moved from one species to another. This is a clear departure from evolution by natural selection, where genes normally move only from parents to progeny within a species. Animals are generally more complex than plants, many having a skeleton, plus the muscles and nervous control necessary for movement. There is thus a far greater likelihood in animals that a biotechnological manipulation will interrupt some important biological pathway, which had earlier been naturally selected to work very efficiently in harmony with the many other such processes. Hence, functioning animals resulting from biotechnological manipulation of genes are far less likely than genetically modified plants.

Breeding Programmes

For many domestic animal species other than dogs, geneticists have developed highly sophisticated mathematical breeding programmes based on estimated breeding values for the products for which the animals are being grown (e.g. meat, wool and milk production). In genetic theory, these estimated breeding values (EBVs) represent average values of the genes for the desired characters. If an animal could be mated to many randomly selected individuals of the other sex in the population, theory expects the EBV value of any trait to equal twice the difference between the average phenotypic value of the resulting progeny and the population mean. Obviously the higher the EBV value is, the more desirable the animal is. To handle the necessary mathematics, simplifying assumptions have been made. These include that correlations among the various traits being simultaneously improved are linear. There is also an expectation that one measurement of a trait will be highly related to subsequent measurements of the same trait if measured again later in life. Rapid progress in the direction of the traits selected has usually occurred. But unexpected and unintended side effects, such as poor reproduction (reduced fitness), have often marred later generations in the breeding programmes.

In contrast, my graduate students and I have developed very effective large-scale breeding programmes for improving guide dogs for the blind, and later for detector dogs, and these have had no deleterious or unintended side effects. As well, over the centuries, hunters, herders and other dog users have selected dogs and created breeds of dogs appropriate for each kind of behavioural work. This phenotypic selection for work has also been very successful. Many people contributed to success, each breeding with those dogs which worked particularly well for them. In the last 150 years, breed clubs and canine associations have flourished. In these, the 'best dogs' were chosen mainly on appearance relative to a 'breed standard', which essentially describes the outward appearance of the 'ideal dog' for each breed. As breed standards displaced working behaviour as the target of selection, working abilities have declined rapidly. Brain tissue is a heavy user of metabolic resources and requires continuing selection to maintain its level of function.

Compared with quantities of products and growth rates or sizes of animals, it is much more difficult to specify and measure behaviour required for a task. Nevertheless, geneticists have started to think about the possibilities of applying sophisticated breeding programmes which work for production traits to behaviour in working dogs. I greatly prefer 'old-fashioned' selection on the working behaviour demonstrated by each individual dog. Our evidence demonstrates that, when properly done, selection on working phenotypes always achieves good results.

Resource Allocation Theory

Was Darwin wrong when he saw that natural selection of animals led to selection responses in populations? And were the founders of population and quantitative genetics wrong when they thought of organisms whose genes produced phenotypes, which were then evaluated and selected? In the historical overview above, selection of animals on their phenotypes seemed to make sense. It still does! So, if there is a problem with breeding plans based on genetic theory, the problem is likely to arise from the shift of focus from whole organisms to genes.

Geneticists currently apply their theory as if environments do not limit the organisms living in them, particularly if they are domestic animals. I have argued that this is why highly productive animals show deleterious side effects as their production increases.

In Beilharz *et al.* (1993), my graduate students and I described in two equations what happens to fitness under natural selection. Equation 1 states an obvious fact. The number of an individual's progeny that get into the next generation (its fitness through reproduction = F) is a product of how often it has young (A), the average litter size (B) and the average survival rate of these progeny to enter the next generation (C).

$$F = A \times B \times C \qquad (1)$$

Equation 2 describes another fact we all take for granted. Animals need food to live. The equation allocates the total energy obtained from the environment (R) to the metabolism going on in the various traits as the animal grows, develops and lives. Resources 'a' go into trait A, 'b' into B, 'c' into C, 'm' is the energy lost because metabolism is not 100% efficient, and 'x', 'y' and 'z' are resources going into traits X, Y and Z which stand for everything else required to keep the animal alive and able to reproduce.

$$R = a + b + c + \ldots + m + \ldots x + y + z \qquad (2)$$

Fitness stops increasing once R reaches its environmental limit and available resources are used as efficiently as possible. This obvious truth was missing in quantitative genetic theory before then. This has led to problems when traditional quantitative genetics was applied to breeding programmes. Until resource allocation theory becomes more widely accepted, the existing problems in breeding programmes based on current genetic theory will continue to occur. What are the consequences of these two equations as animals adapt to their environment? Beilharz and Nitter (1998) have set out the details of the resource allocation theory derived from these equations.

The best adapted animals are using the resources of their environment in the most efficient way to achieve highest fitness. Unless the resources of their environment increase, animals are prevented from changing from the optimum values their phenotypic traits currently have, given the constraint that R has reached its maximum value. What are these optimum 'intermediate' values?

For reproductive component traits A, B and C, the following analogy applies. Which three numbers summing to the constraint (say 12) have the highest product?

The answer is 4, 4 and 4 (as $4 \times 4 \times 4 = 64$). Every other set of three numbers limited to a total of 12 has a lower product ($5 \times 4 \times 3 = 60$, $6 \times 3 \times 3 = 54$, and so on). Note that in such a constrained situation with product at its maximum, increasing any trait can only reduce the product. And it is the product which is being maximized by natural selection.

For other traits, say X (e.g. hair cover), there will be selection for some hair as with no hair the animal dies from cold (fitness = 0). If hair cover becomes more dense than necessary, the animal may have depressed reproduction because of heat. But even in the absence of a problem with heat, if x is higher than necessary, the proportion of R available to maximize fitness (a + b + c) is reduced and F decreases. So, natural selection on fitness restricts all traits with an effect on fitness to intermediate 'optimum' values.

This is an empirical fact already stated by Crow (1986). Crow considered this effect of natural selection to be a paradox. He discussed possible genetic explanations for why extreme phenotypes were being culled. Long before Crow, Fisher (1958 – first published 1929) had postulated another genetic explanation for the fact that, in adapted populations, fitness usually does not rise. His 'fundamental theorem' is that increase in fitness is directly related to available genetic variation in the trait fitness. This implies that there are special genes for the trait fitness. But fitness is the end result of all traits that affect reproduction or survival (that is, traits necessary for life). There is no viable population in which there is no variation in every trait important to life.

Crow also commented that selection for intermediate values, in each of many traits, would allow populations to rapidly track any environmental change. Each trait would move rapidly to its new 'optimum' value. What I have done here is to substitute an obvious explanation from resource allocation theory for speculative genetic explanations. Nevertheless, I remain deeply indebted to Crow for his clear description of what natural selection produces.

One very important consequence follows from the above. In a stable environment, once animals are adapted to their environment, natural selection prevents phenotypes changing from optimum values. The fossil record shows long periods of tranquillity with unchanging fossils until an extinction occurs, after which survivors change rapidly to fill all niches now available after the catastrophe causing the extinction. Hence, for most of evolutionary time natural selection has prevented evolutionary changes, because each environmental niche was limited by its resources. Catastrophes following cosmic, geological or climatic events cause extinctions of many species when habitat is destroyed. When the Earth settles down again, the survivors, now unconstrained by environment, can rapidly respond to natural selection, which adapts them to one or many newly available niches. Until environmental niches become limiting again, genetic changes are very quick, on an evolutionary time scale. Inevitably, natural selection will again cull unsuitable 'extreme' phenotypes as environmental niches start exerting constraints. Genetic changes do express themselves when unused resources are available. But they cannot change phenotypes markedly in adapted animals limited by their environments. This fact has been sidelined in the advance of biotechnology and modern animal breeding. It suggests that many goals of breeding programmes based on traditional quantitative genetics may not be achievable.

There is one other important consequence. If the new environmental niche does not require change in any feature of an organism, this feature will not change from its previously optimized form. When mammals replaced dinosaurs, basic mammalian features like temperature control remained unchanged in the mammalian radiation. Furthermore, mankind has the same efficient early embryological development as many other animals including even fish, because the liquid environment in which early embryological development takes place has not changed. Our own research has provided several examples of artificial selection demonstrating that only the specific trait selected has changed. Wool sheep, selected for fine fibre on midside samples, have fine fibres in the midside, and far less elsewhere (Stadler and Gillies, 1994). Mice selected at first parity for litter size have very large first litters, but hardly any change in later litters (Luxford *et al.*, 1990).

Taking note of environmental resources, and specifically that available resources may limit the phenotypes animals can achieve, should be incorporated into breeding programmes aimed at genetic improvement of animals. On farms, as in breeding working dogs, we must evaluate whole animals on the basis of their phenotypes. Looking at genomes, estimating breeding values and searching for quantitative trait loci (spots on chromosomes with large effects on quantitative traits) is looking at the wrong level. It is like analysing the single threads in the cloth to determine whether we are looking at a pair of ladies' jeans or men's jeans. When we work with whole animals, it becomes clear that they are limited by their environment and the food available. We should then also see clearly that our farms cannot sustain animals that are 'two standard deviations better than our present stock'.

What is the Behaviour Required by Working Dogs

A short discussion of different types of behaviours and how genes control behaviour is useful, before discussing how dogs can use their behaviour for our benefit. Can we visualize how genes and the environment combine to achieve behaviours and other traits? How do traits interacting with fitness in our equations express themselves in real animals in the real world? Genes work indirectly. How do genes make characters? One idea is very important! Genes usually contribute to a system moving along a track in time, a developmental path.

Very rarely do we recognize the action of single genes on their own. Take as an example something we can visualize, a gene for horns in cattle or sheep. Many genes together influence the formation of horns. But genes are not the only causes. Sex hormones contribute to horns taking a shape in males different from the shape developed in females. Similarly, if particular nutrients are in short supply in the food, stunted horns may result. So, development of horns occurs in an external environment which needs to be adequate, in a physiological environment in which hormones influence sexual differences and, of course, within a system in which many genes are cooperatively contributing structures and control of the developmental sequence. The same considerations apply to phenotypic traits in dogs such as size and shape of ears.

Every gene can mutate from one allelic form to another, though it does so rarely. In a population in which horns are reliably formed in every individual, the alleles present at each of the structural and controlling genes contributing to full horn formation are called the 'wild type' or 'normal' alleles. If random mutations occur at any one of the structural or controlling genes, the development of complete horns can be interfered with by the new 'mutant' alleles. A reaction can be slowed down or diverted from the normal path resulting in abnormal horns. Or the process may be interrupted completely, so that no horns form at all.

Alleles for polledness, the complete absence of horn production, could be alleles of any one of several genes that must work together to achieve the final product. Cattle breeders do not care which gene is responsible, as long as they get the phenotypic result, no horns. But, you cannot use the wild-type form of that allele of itself to create horn production in a test tube, unless you also have wild-type alleles of all the other genes involved in the process. (The news media often proclaim that a 'new gene for medical problem M' has been discovered. Recognizing such a gene may allow one to repair a particular interruption to the pathway to health. But it does not guarantee that an interruption at a different locus affecting the path to healthy development will not appear.) Genes for floppy ears in dogs interrupt the full process that made the ears of a wolf stand up fully and be highly mobile so that even faint sound from any direction in which the ears are pointed could be picked up (also see Chapter 6 in this volume).

Genes cooperate in sequences in time, either by providing material substrates to the sequence or contributing to the timing of the sequence. A gene by itself is an unhelpful concept in complex forms of life. We should think of a living organism as a sequence in time, starting from the fertilized zygote and ending with chemical decay of the formerly living phenotype, when the biochemical reactions that maintain life stop at death. We must now try to convert the theoretical notions of reproductive and other fitness traits into concepts that help us visualize the life of an organism when this is seen as a developmental sequence of biochemical reactions from fertilization to death.

We can begin by saying that length of life (strictly, number of parities) is an important character. Natural selection must have brought it into an optimal relationship with all other reproductive and survival traits.

In life, metabolic decisions about where resources must go will be taken moment by moment. During development of a bird or mammal embryo, important metabolic decisions are made in the mother, either when provisioning the eggs she will lay, or nurturing the fetus inside the uterus. Once young are born or hatched, parents usually still help for some time. But more and more the young must fend for themselves. In other words the external environment starts to challenge the individual young growing animals and soon they must obtain by themselves all necessary resources for further development and survival. As they approach maturity, there is a period where animals are still growing but reproduction is possible and will take place if resources are plentiful. Otherwise the animal delays reproduction for another season until it is more mature and then starts to reproduce. Thereafter, the animal essentially spends its time reproducing, and then recovering from the metabolic drain of putting resources into young.

What natural selection does is to remove, by death, all individuals in which at any moment metabolic decisions have left an animal unable to cope with the

Fig. 10.1. A detector dog at work.

current challenge. It also discriminates against parents less able to compete for food and thus having fewer young. Similarly, the simple loss to a rival of an opportunity to mate is another way for natural selection to penalize an organism. There is thus continual balancing of everything biological that you can think of. Some examples of unsuccessful evolutionary strategies are:

Fig. 10.2. Detector dog coming off plane after work.

Fig. 10.3. A potential guide dog showing a certain amount of fear towards the handler.

1. Litter size is too large when there is no mechanism to reduce litter size, and mother dies.
2. Reproduction starts too early in life with stress of reproduction too great to maintain the still growing animal through the next difficult season.
3. Metabolism does not use all available resources, causing the animal to have relatively few young.
4. Young animals walking up to 'unfamiliar beings' when they emerge from their protected nests.
5. Birds making nests on cliff ledges, before young have an 'instinctive' avoidance of cliff edges (in humans we call this 'fear of heights').
6. Animals having to learn escape responses by experience, rather than using 'instinctive' fleeing or freezing reactions.

You can imagine hundreds of other possible deviations from efficient developmental paths, which will be penalized by natural selection. Although each metabolic decision is made independently at each moment, natural selection in every generation rewards genetic changes that anticipate regularly appearing challenges in the environmental niche along the lifetime developmental path. This means that adapted organisms already inherit pre-programmed developmental paths in which resource needs for an efficient path are anticipated and prepared for. Examples are laying down of reserves of fat for anticipated future

Fig. 10.4. Detector dog coming off a boat.

needs, giving young animals emerging from safe nests an instinctive fear of unfamiliar beings and giving nestling birds on cliff ledges a similar 'fear of heights'.

As the environment selects organisms in development programmes, what roles do genes play? Every development programme is guided by the information in the particular set of genes put into the zygote by both parents. Whether genes prosper depends on how well the developmental path launched and guided by the genes of each zygote or clone is able to handle the challenges of the local environment. Developmental paths of organisms are continually under pressure, metabolically, to keep the organism alive. In the products of evolution which nature has provided, there is very little scope for an individual animal, on its development path, to 'take it easy'. This is why animal breeders are unlikely to change animals usefully, without paying a biological cost, when they add a new feature or exaggerate an existing one. Breeders' wishes involve costs.

We can summarize the consequences of resource allocation theory as follows. The model of fitness as stated in our equations is a simplification. A living organism should be seen as a developmental path from the fertilized zygote to death. Genes influence this path. Natural selection eliminates all organisms that at any moment have insufficient metabolic resources to cope with the next challenge from the environment and discriminates against all those not using available resources efficiently. This in turn selects for the putting down of reserves for predictable changes in resource needs, such as seasonal or pregnancy-related extra efforts. Organisms inherit programmed lifetime developmental paths. These programmes include mechanisms for adjusting reproductive effort to the environment. In each environmental niche there are optimal sizes and optimal developmental paths. Natural selection uses the processes of genetic canalization and genetic assimilation to achieve optimal lifetime paths. Deviations from optimal sizes and optimal developmental paths incur metabolic costs. These typically lead to reductions in reproduction and longevity. 'Premature' ageing is a consequence of above-optimal resource use, often early in the lifetime path.

Behaviour is classified into several categories. Reflexes, instinctive behaviour and learned behaviour are behaviours that are important to working dogs. Reflexes are responses to stimuli before the animal becomes aware or conscious of the stimulus. Instinctive behaviour is that behaviour which occurs when a stimulus is encountered for the first time. Both reflexes and instinctive behaviour are the direct result of natural selection in the evolution of animals. In the developmental path of an adapted animal, the animal must be able to deal with a stimulus the first time it experiences it. Hence, both reflexes and instinctive behaviour need to be essentially correct the first time the appropriate stimulus situation is experienced. Reflexes include responses to gravity or to pressure on the body and withdrawal responses to hot or painful stimuli. Instinctive behaviour includes avoidance of cliff edges and unfamiliar beings outside the familiar nest, the sexual reactions by males to females and vice versa and reactions between mothers and newborn young. Any animal that reacts falsely is left behind by natural selection. So reflexes and instinctive behaviour in normal healthy animals are dependable and appropriately motivated. The level of motivation also results from natural selection. Clearly, instinctive behaviour and motivational level of working animals will also respond to artificial selection by humans and they are very important to successful breeding programmes.

Learned behaviour relies on making associations between stimuli and behavioural responses. This process is known as conditioning. Classical conditioning is the provision of a neutral stimulus, like a sound, shortly before providing food. Provision of food automatically stimulates the eating response and all its motivations. If the sound reliably predicts the arrival of food, the sound becomes the conditioned stimulus which by itself releases the motivations associated with eating, and the eating itself if food then appears. Repeatable signals preceding the regular feeding of animals in zoos or livestock in droughts become strong conditioned stimuli.

Instrumental conditioning (see also Chapter 8 in this volume) is the process where a stimulus places the animal under some stress. If the reaction of the animal relieves the stress, the animal associates its reaction with removal of stress and, provided signals are clear and consistent, behaviour will become shaped such that stress is minimized. Such instrumental conditioning is very important in training dogs and horses into precise behavioural responses. For many working dog situations, both classical and instrumental conditioning play a part. Breeding programmes are likely to be less affected by learned behaviour than by instinctive behaviours. Selective breeding will readily change motivation levels associated with instinctive behaviour. It is difficult to see how selective breeding will affect the speed by which dogs make associations.

Successful Breeding Programmes

The most important item in every practical breeding programme for working dogs is to define very precisely the phenotype that working dogs must have. This depends on the job that the dog is to do. Persons able to make dogs work well are essential in defining the phenotype to be aimed at. Understanding dogs, their behaviours, their abilities and how they interact with people, is much more important than any genetic considerations. After defining the desired phenotype, one selects as breeders those animals with the best working phenotypes and breeds with them. That is the outline of the task. Generally, when a trait has not been selected in any population, there will be a selection response for 10–20 generations. The most spectacular gain will occur in early generations. Whether the programme is successful depends on the details put in place to achieve the breeding, testing, training and recording of traits.

Long ago we did this for Australia's major guide dog breeding programme and, more recently, for the Australian Customs' detector dog programme, which is now reputed to have the world's best detector dogs. In both cases we followed a set of six steps as follows:

1. We need to understand the background and aims of the breeding programme. The better we understand the background and aims, the more appropriate will be our breeding programme.

2. In every breeding programme, the above aims must translate into changing particular phenotypic traits of animals. There are two important considerations in defining the traits: (i) are the traits important in meeting the aims? Ignore unimportant traits; (ii) how are the traits to be measured? Unless you can describe a trait on a dog by a number, you cannot say which animal is to be preferred. The traits most difficult to assess are often the most important: instinctive behavioural tendencies, concentration on a job, lack of fear. Here you need experience with dogs, or the advice of someone else with experience, to rate dogs subjectively on a 5- or 7-point scale, where the middle value describes the average dog. For working dogs we want all the important traits to be expressed in a healthy animal, over a long working life. We breed from a long-lived successful worker when this can be done.

3. Find the genetic information for each important, measurable trait. You need for each one: heritability (h^2) and degree of heterosis or hybrid vigour; and for each pair of traits: whether they are genetically correlated. You need to select only one of two highly positively correlated traits as the other will follow automatically, at least for some time. Negatively correlated traits must both be selected simultaneously.

It may be difficult to find genetic parameters in the literature for traits over a lifetime. For traits using metabolic resources heavily, early high production may shorten lifetime. We need to select the whole lifetime pattern of expression of the trait which our animals are to achieve.

4. Choose breeding strategies. Use selection on heritable traits and outbreed or crossbreed traits showing hybrid vigour. Traits already selected strongly, either artificially or naturally, will not respond much to further selection. For traits not previously selected, expect a good response to selection, at least initially.

5. Synthesize all the separate strategies into one overall breeding programme. The task is to combine selection for traits of high heritability with crossbreeding for traits with hybrid vigour.

We want to find those animals that have the best combination of individual traits to breed with. Independent culling levels, in which we cull dogs poor in any of the characters we want, is appropriate for heritable traits. Ensure that all heritable traits you want are selected appropriately and make those matings which provide the outbreeding necessary for the traits showing hybrid vigour. Insure against inbreeding decline becoming a problem in future by keeping the population base wide rather than narrow. Cooperate with breeders following similar goals in similar environments and periodically bring in unrelated stock.

6. Check the overall plan for feasibility and for its effectiveness in meeting breeding goals.

In the two Australian dog breeding programmes, we worked very closely with guide-dog trainers and customs dog handlers to define the traits required in the respective programmes. We found no useful heterosis in either programme. In our guide dog research, Mike Goddard bred hybrids between labradors and three other breeds of similar size. The hybrids were slightly more wary, a sign of positive hybrid vigour in natural evolution, but opposed to the complete freedom of fear and distraction that we were looking for as the most important trait for working guide dogs. In detector dogs, Kath Champness and John Vandeloo from the customs service devised a chase–retrieve test for finding an article with an odour in increasingly difficult places. They rated the dogs subjectively on several aspects, including their ability and willingness to find, retrieve and hold on to the article.

The success of the guide dog breeding programme is determined largely by finding healthy dogs (guaranteeing a long working life), which are free of fear and distraction. This requires selection for instinctive behaviours and appropriate motivation levels. Even if we cannot yet describe exactly the several traits involved in being free of fear and distraction, the dogs that demonstrate the desired behaviour must have had the genes we want in the rearing and training environment in which they grew up. When we work with phenotypic results we make progress, just as natural selection does. It does not matter that the psychology of the dogs is not yet known in detail.

For drug detector dogs we used surplus dogs from the guide dog programme, initially those that were too exuberant. But almost all these dogs were sufficiently free of fear and distraction to be trained as detector dogs. One major criterion for their selection was that they should be driven to search for and retrieve an item with a specific odour, with the dog then being rewarded by the handler with a game of tug-of-war. This complex trait involves high instinctive motivation for searching and retrieving, the willingness to keep on doing so without tiring and also instinctive motivation for playing with the handler. Ability to make associations with specific odours is also required. Again, those dogs that can do this are the dogs we want. They have the correct genes for the appropriate instinctive motivations and enough ability to make correct associations. As this character is unlikely to have been selected heavily among labradors, one can confidently predict a substantial selection response. This was indeed the case.

Concluding Remarks

Although neglected in the current theory of genetic improvement and biotechnology, the dependence of animals on the resources of the environment is very important. In this chapter I have discussed why environmental resources are now not taken seriously, even though Darwin's discovery required recognizing that the most efficient users of resources were the successful animals selected in nature. Essentially it is the change of focus from whole animals to genes, and recently to the molecular structure of genes, which prevents us seeing that the phenotypes of whole animals are limited by their current environmental resources. The resource allocation theory explains how and why environmental resources limit phenotypes for most of the time. The exceptions are when new resources are becoming available or when former users of resources have disappeared as a result of natural or man-made causes.

I also described how genes should be understood as contributing to the lifetime developmental path of individual animals. It is such lifetime phenotypes which natural selection optimizes for fitness and which breeders should be trying to change in their quest for more commercial products. Looking at developing phenotypes also helps us to see instinctive and reflex behaviours being selected by nature and how breeders can change behaviours to achieve better working dogs.

Successful breeding of working dogs requires an intimate knowledge of the working environment and the behaviours that each dog must have in order to become good at its work. Breeding with the dogs that are the best workers is using artificial selection together with natural selection on health, longevity and reproduction in the same positive way as evolution has used natural selection on fitness in each environmental niche. Attempting to use existing programmes for calculating EBVs for different 'working traits', or finding genes for individual traits, will merely introduce unnecessary complexity, dependence on mathematical simplifications and delay into the programme. Two very successful Australian breeding programmes for dogs suitable as guides for the blind and detector dogs demonstrate that phenotypic selection for precisely the work needed is highly successful.

References

Beilharz, R.G. and Nitter, G. (1998) The missing E: the role of environment in evolution and animal breeding. *Journal of Animal Breeding and Genetics* 115, 439–453.

Beilharz, R.G., Luxford, B.G. and Wilkinson, J.L. (1993) Quantitative genetics and evolution: is our understanding of genetics sufficient to understand evolution? *Journal of Animal Breeding and Genetics* 110, 161–170.

Crow, J.F. (1986) *Basic Concepts in Population, Quantitative and Evolutionary Genetics*. Freeman, New York.

Darwin, C.R. (1859) *The Origin of Species*. John Murray, London (New American Library, Mentor edition, 1958).

Fisher, R.A. (1958 – first published 1929) *The Genetical Theory of Natural Selection*, 2nd revised edn. Dover, New York.

Luxford, B.G., Buis, R.C. and Beilharz, R.G. (1990) Lifetime reproductive performance of lines of mice after long term selection for first parity litter size at birth. *Journal of Animal Breeding and Genetics* 107, 188–195.

Mendel, G. (1866) Experiments in plant-hybridisation. In: Sinnott, S.W., Dunn, L.C. and Dobzhansky, T. (1950) *Principles of Genetics*, 4th edn. McGraw Hill, New York, pp. 463–493. (Translated from German by the Royal Horticultural Society of London.)

Stadler, W. and Gillies, R.I. (1994) A case study for the use of mid-side samples to predict the mean fibre diameter of classed lines. *Journal of Wool Technology Sheep Breeding* 42, 319–326.

11 Individual Differences in Behaviour – Dog Personality

Kenth Svartberg

Introduction

Dogs are highly capable of adapting to new environments and learning to perform different behaviour in certain situations (see Chapter 8 in this volume). However, some aspects of a dog's behaviour might have limited plasticity. In fact, if you observe dogs and are focused on behavioural stability, you might find the dog quite consistent in a range of situations. For example, the dog's strategy when meeting unfamiliar persons may be very similar in different contexts and over long periods of time. A dog that shows signs of fear when exposed to loud and strange noise may still, several years later, show similar tendencies in such situations. Other examples are a dog's typical tendency to get excited or to be aggressive. Such stable dispositions create what could be called the behavioural style of a dog, which has also been referred to as temperament, individuality, coping style, behavioural syndromes and, more lately, as animal personality.

Traits as Complexes of Behaviour

Central for the issue of individual differences in behaviour, which I will refer to as personality in this chapter, is *trait*. If you observe the behaviour of a dog you will probably find that some behaviours often come together. For example, the dog that snarls when meeting other dogs will probably also raise its tail, lower the head, stare towards the other dog, bare its teeth, and perhaps also lunge towards the dog. Such a 'package' of behavioural reactions may be labelled a behavioural trait – a hypothetical construct with which it is possible to describe the behaviour of an individual, as well as differences in behaviour between individuals. In everyday terms we probably would like to call this trait hostility, or aggressiveness. If you have the possibility to observe several dogs in a similar dog meeting situation it is very likely that you will find that dogs differ in regard to this trait. Some dogs will display these behaviours early in the meeting situation and with high intensity, other dogs will show some of the above-mentioned reactions, but not all and not very intensely, and still others may not show any sign of this trait.

From Behaviour Traits to Personalities

Is behaviour the same as personality, and are behavioural traits similar to personality traits? The study of personality in humans is closely related to the assessment of feelings, thoughts and beliefs. Within the study of animal behaviour, internal processes – such as feelings and thoughts – have been considered unobservable or even scientifically irrelevant. Researchers have striven towards explanations of animal behaviour in the simplest possible way, in accordance with Lloyd Morgan's Canon, and have avoided the use of unobservable events such as emotions and intentions as behavioural explanations. As a result of this, it seems that scientists interested in individual differences in animal behaviour also have avoided the concept of personality because of fear of anthropomorphism. However, besides feelings and thoughts, personality in humans also includes an issue that is possible to study in animals – behaviour. Personality traits can be described as dispositional factors that regularly and persistently determine behaviour in many different types of situations. Thus, an individual's personality can be inferred from the individual's behaviour. This makes the study of animal personality no different from any other studies of animal behaviour. The label of a suggested trait – such as 'fearfulness' or 'aggressiveness' – may only be a short description for a disposition of the individual to act in a certain way, and does not necessarily imply the existence of any feelings or thoughts. However, current research gives evidence for the existence of different basic emotions in animals, as well as cognitive processing that may be analogous to human thinking at a less complex level. It is probable that a dog that waves its tail and steadily focuses on the owner when he or she is going to throw a tennis ball really is full of expectancy and is anticipating the coming game, even though both expectancy and anticipation are private for the dog and not observable. Assumptions of emotions and internal processing in animals give even more relevance for the study of personality in animals. For example, rats that live in an unpredictable environment seem to be more pessimistic – having negative expectations about unknown events – than rats with a more stable life situation (Harding *et al.*, 2004). Development of new experimental designs has recently yielded tools in understanding cognitive skills in dogs, such as word learning, numerical competence and use of social cues to get rewards, which may help us to unravel individual differences in cognition (see Chapter 12 in this volume). However, there is still a need for highlighting the

Fig. 11.1. Behavioural reactions that are stable over time and across contexts can be assumed to be expressions of the dog's personality.

risk of anthropomorphism in assessing personality traits in species like the dog. Personality assessments in animals should primarily be based on behavioural observations, and not on assumptions of thoughts and feelings.

So, behavioural observations may be useful when we want to know something about the dog's personality. But are all behavioural reactions expressions of the individual's personality – is a behavioural trait the same as a personality trait? For example, can we be sure that the snarling border collie in Fig. 11.1 has an aggressive personality? Here we may use the definition of personality traits again: 'dispositional factors that regularly and persistently determine behaviour in many different types of situations'. In this definition, two aspects of stability are included – stability over time and stability across situations. When taking the first of these criteria of personality into account, it is possible that the aggressive reaction of the border collie is specific for this specific occasion. A similar situation the next day, month or year may give a different reaction. The reason for such time-dependent specificity may be that this reaction is sensitive to experience, which gives that the dog's reaction may change every time it is exposed to a certain situation. Thus, a behaviour that is easily changed by training should not be seen as an expression of the dog's personality (even though this might tell us something about the dog's general trainability, which can be part of a dog's personality). Change in behaviour from one time to another may also be due to maturation. For example, a male dog's non-response to a bitch in heat may give reason to assume that this dog has a low 'sex drive'. However, this assumption may be very misleading if the assessment is made at a young age. A test later on, after sexual maturation, could give a totally different picture. Similarly, behaviour reactions observed at a very young age may correspond poorly with reactions observed in a similar situation later on in life. Therefore, it is important to take maturation into account when assessing a dog's personality – behavioural strategies are a part of the personality first when it is to at least some degree temporarily stable.

The second criterion from the definition of personality was stability across situations. A single observation of a dog may prove to be highly situation-specific, and may not be relevant at all in other situations, for example in a test. Even though the dog's behaviour is very similar every time it is exposed to a certain stimulus situation, it may be specific for this particular stimulus situation. For example, your dog shows avoidance reactions to Bill, who is a friend of yours. Through several meetings with Bill you know that this reaction is stable over time and exposures. Can this information be used to assess this dog as avoidant to humans, or at least men? I believe that your answer is 'no', since we have no information about the dog's reaction towards other persons. It is possible that this dog may show no avoidance reactions to other persons, which means that an assessment based on the first information should have been misleading. Similarly, a dog that is superior to other dogs in learning one task, but not another (where it is rather poor) should not be regarded as highly trainable or intelligent. In the definition of personality some degree of generality is inherent, which also tells something about one fundamental issue in the concept of personality – making predictions of the individual's behaviour from one situation to another similar situation. A reaction that is too specific, even though it is stable over time, is rather useless in prediction of behaviour and may not be a relevant measurement of the dog's personality.

Thus, a glimpse of a dog's behaviour may say something about its personality *if* the reaction is stable in two regards – stable over time and across similar situations. A behavioural trait can be referred to as a personality trait after positive tests according to these criteria; a personality trait is a hypothetical construct useful to describe the individual's *typical* behaviour. However, to expect that dogs or other living organisms are stable in the sense that they always show the same behavioural response is misleading. On the contrary, a dog that always behaves in a certain way – say, always waving its tail – should probably best be regarded as pathological, rather than having a stable personality. Personality in a dog should always be seen as an interaction between the dog and the environment, and assessed in a context – 'if this happens, or in this type of situation, the dog usually behaves in this way'.

One last issue regarding personality traits is how the trait is best described, and the distribution of different expressions in a population, say, a breed. A common assumption is that behavioural traits can be described linearly and are normally distributed; an individual can be assessed according to a dimension, and more individuals are assessed as intermediates than extremes. Most personality traits in both humans and animals seem to have this characteristic, even though the distribution is not always normally distributed (for example, skewed towards one end-point of the scale). In contrast to the assumption of linearity and normal distribution, there are some suggestions of traits that have a different characteristic. Within animal personality, the most well-known example is coping style, which has been studied in species like rats, mice, great tits and pigs (Koolhaas *et al.*, 1999). According to the theory behind this construct, some individuals have a disposition to cope with stressful events with activity and aggressiveness. These animals are also more prone to create habits, and, thus, less flexible when circumstances change in a familiar situation. Besides this coping style there is another type, which is characterized by a more passive strategy when stressed: low aggressiveness, prone to be inhibited and have a better capacity to adapt to changes in the environment. The distribution of this trait has been assumed to be bimodal – two separate and extreme types with no intermediate individuals. However, this can be seen as an exception, which also has been questioned (Jensen *et al.*, 1995). A safer stand-point is to assume that individuals differ in personality according to dimensions from low to high (or low intense to high, seldom to often, hard to elicit to easily elicited, etc.) regarding the behavioural reactions that the trait refers to. This means that personality descriptions often are relative rather than exact: 'dog A is typically more fearful than dog B'. This is contrasted with everyday descriptions of personality where references to types are common. For example, 'his dog is aggressive' or 'my dog is playful'. Because of what has been mentioned regarding distribution of personality traits above – types rarely exist – such labelling is misguiding.

Why Study Personality in Dogs?

There has been an increase in interest in the study of animal personality, as well as in personality in dogs, during recent decades. It seems that this change is driven by a parallel increase in interest in other areas. One of them is animal welfare. If different individuals appraise threats or other stressful events differently, some individuals will be better able than others to cope with a certain life situation. One example in dogs is the capacity to cope with temporary separation from family members, or social isolation. To ensure that dogs may cope with this situation, which is a very common for pet dogs, it is important to take each dog's personality into account. Some dogs may be more easily trained to cope with loneliness, whereas others may suffer more in such a situation.

Another issue is prediction of behaviour. Knowledge of future ways of acting in different situations is valuable in selection of potential working dogs, such as guide dogs, dogs that are used for search tasks (explosives, drugs, etc.), guard dogs, hunting dogs and herding dogs. Behavioural signs in a dog that predict success or failure before the dog is trained, or in the early phases of the training period, bring great advantages. Time and money can be saved, and the welfare for dogs and trainers may be improved. Behavioural prediction in dogs may also be important for pet dogs. Early signs in the puppy may help the breeder to match the dog with an appropriate owner. Furthermore, in several countries there are extensive rescue dog programmes. Dogs that otherwise would be put down may be transferred to a new home after a stay in a shelter. An assessment of the dog's typical behaviour in certain situations when it is in the shelter may increase the chance of a good match between the dog and the new home. Behavioural problems may be avoided, or at least diminished, by taking preventive steps. Furthermore, the value of early signs of behaviour that may cause problems for the owner, the surroundings and/or the dog itself may be extended to all pet dogs. Knowledge of the young dog's typical behaviour may help the owner to take appropriate steps, such as training programmes, that make life more pleasant for all parties.

A third issue is evolution or, in the case of the dog, domestication. What traits are favoured during selection, and why? In dog breeding, there is – consciously and unconsciously – a selection for wanted traits and against unwanted traits. If we assume that these traits have a genetic base, the type of selection that dominates will decide the typical behaviour of the dogs in future generations. Thus, methods that are useful in assessing the personality in breeding dogs as well as the offspring are of great interest for successful directed selection. For example, standardized behavioural tests have been used as tools in breeding programmes in breed clubs and in selection of working dogs. In addition, it is relevant to understand how other, more unconscious, selection criteria influence the ongoing domestication of dogs. Dog personalities differ in their adaptiveness in different life situations and contexts. A certain dog personality might be highly adaptive in one setting, whereas the same dog might give another owner problems in everyday life due to its typical behaviour. One example of this type of conflict is seen in Fig. 11.2. A shyness–boldness dimension has been detected in dogs in a standardized behavioural test (Svartberg and Forkman, 2002). This trait is related to everyday life – playful, explorative and fearless behaviour in both social and non-social situations – as well as to success in working dog trials (Svartberg, 2002, 2005). Depending on the goals of the owner and the breeder, different dogs will probably be favoured during selection. The bold dog might be favoured if the

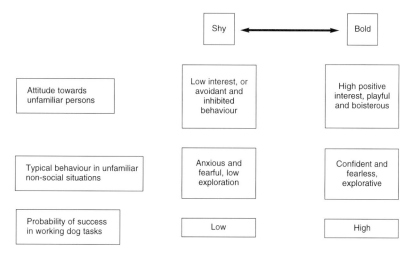

Fig. 11.2. A certain dog personality might be highly adaptive in one setting, whereas the same dog might give another owner problems in everyday life due to its typical behaviour.

major goal is some kind of working performance, whereas less bold dogs, although not shy, may be more easy to handle for pet dog owners. This assumption is supported by correlations between the typical use of parents in different breeds, and breed-typical personality (Svartberg, 2006). Breeds where breeding dogs have a high number of merits from working dog trials are in general more playful than breeds with parents less often used as working dogs. Furthermore, 'show breeds' are shyer than breeds where show merits seem to be less important.

Scientific Methods of Assessing Dog Personality

Where should personality be assessed?

What do scientists do when their aim is to assess personality in dogs? Different methods are used, and there are some factors that influence the choice of method. For example, some aspects of the dog's personality may only be possible to assess over a long period of observation and in certain situations. If we assume that dominance, as an example, is a possible personality trait in dogs, a single behavioural test will probably not give us the whole picture of the dog's tendency to dominate or show submissiveness towards persons or other dogs. Dominance relationships are formed within the social group over some periods of time, which makes a behavioural test inappropriate in this regard. One way to get this information is to ask a person who is well acquainted with the dog about the dog's typical behaviour with other well-known dogs and with familiar persons. An alternative way could be to get access to the dog for a certain period of time, and study the dog's social behaviour in a controlled environment. Another issue could be the degree of generality. If, for example, the aim of a study is to compare the typical behaviour between several dog breeds, the personality of a large number of individual dogs has to be assessed. This limits the sampling method. To observe a sufficient number of dogs per breed in a behavioural test may be practically impossible, which justifies other methods.

Issues like these have led to researchers using different approaches in the scientific study of dog personality. Generally, three methods of collecting data on dog personality can be defined: (i) observation of the dog's behaviour in its normal environment; (ii) behavioural tests; and (iii) information from persons who are familiar with the dog, for example questionnaires sent to dog owners. Behavioural tests have been used in order to assess traits like greeting behaviour in meetings with unfamiliar persons, fearfulness towards different sudden or novel stimuli, and aggressiveness towards unfamiliar and threatening persons. A test situation, however, can be seen as a novel situation in itself, which limits its usefulness. The dog's typical behaviour may be masked by its reaction towards the test situation, which sometimes makes it necessary to use other sampling methods. One such commonly used method is questionnaires in which the dog owner rates the dog's behaviour or level of some predefined personality traits. This methodology may capture such aspects of personality as the dog's typical social behaviour within the family, tendency to cooperate with known persons, and behaviour when left at home. Thus, behavioural tests may be useful to get information about some aspects of a dog's personality, whereas other methods, such as questionnaires, may be necessary in order to gather knowledge of other aspects of the dog's typical behaviour. A well-constructed questionnaire and large sample sizes may, at least partly, compensate for the bias that the large number of observers (dog owners) gives.

Measuring personality

Besides the issue of sampling method, another question concerns how to assess the dog's personality. Personality in animals may be assessed at two different levels – behavioural observations according to strict objective criteria and with subjective assessment. The first of these levels concerns what is the most common method in ethological studies in general. The behavioural reactions are rated according to strict objective criteria – for example, number, frequency, duration and/or latency. This method has been used in theoretical approaches, where behaviour measurements are used as indicators for a suggested personality type (for example, the time until an intruder is attacked by a resident may be used as a base for categorizing the animal as 'non-aggressive' or 'aggressive') or for the magnitude of a trait (for example, the number of threat behaviours observed can be used to assess the animal's degree of 'aggressiveness'). Objective measures of behaviour have also been used in more empirical and exploratory approaches, where clusters of correlated behavioural variables that presumably represent personality traits are searched for by the use of factor analysis or other multivariate analysis methods.

Animal personality may also be assessed at a more comprehensive level, where the observer subjectively rates the individual according to predefined traits. Observers, who assess the personality by observing it in several situations, are here used as data recording instruments. Commonly, the animal is described according to adjectives, such as 'curious', 'motherly', 'playful' and 'understanding', on a linear scale. This method provides a higher level of description, and may capture the overall pattern of an individual's behaviour that remains elusive when discrete events are measured. Studies on human personality have shown that descriptions of behaviour at a more general level may be more predictive than specific measures of behaviour (e.g. Funder and Colwin, 1991). However, direct assessment of personality is sensitive for subjective interpretations, and may easily be biased by the observer. Therefore, the accuracy of this method rests on the use of several independent observers, together with high criteria of inter-observer agreement for the suggested traits.

Besides these two general methods of assessing animal personality, there is one additional approach that is commonly used that may be seen as a 'middle way' – behavioural rating scales. In this method, the animal's behaviour is rated in a specific situation (in contrast to subjective rating, where the overall tendencies are assessed) according to a predefined scale with a number of steps. For example, aggressive behaviour in dogs has been rated according to a five-point scale (Netto and Planta, 1997): 1 – no aggression observed; 2 – growling and/or barking; 3 – baring the teeth; 4 – snapping; and 5 – biting and/or attacking with bite intention. Some additional assumptions are made when rating scales are used, compared to when using strict objective criteria. For example, that several behavioural reactions (such as growling, barking and snapping) are associated with the same behavioural category (aggressiveness). In spite of this, rating scales seem to be useful in dog personality studies, and, furthermore, are easy to use and therefore a common method in applied settings.

Personality Traits in Dogs

Perhaps the most intriguing question is what personality traits are to be found in dogs? This question is not easy to answer, however. In studies where stable aspects of dog behaviour have been in focus, a range of traits have been proposed, but only a few of these have been tested for stability over time and across situations. Thus, we have relatively poor knowledge of the stability of a number of proposed traits, as well as of their relevance in different situations. There are some traits in the literature of dog personality that are more commonly described than others, and more often tested for stability. The two most widely suggested are fearfulness and aggressiveness, which I will describe in more detail here.

Fearfulness

'Fearfulness' is probably the most studied trait in animals, and the domestic dog is no exception. In dogs, there are several behavioural reactions that are commonly associated with fearfulness. Examples are avoidance behaviour, flight behaviour, low body posture with low tail and ears, trembling, salivating and vocalization, such as yelping and screaming. Behavioural reactions associated with fearfulness have also been regarded as expressions of other traits, which may be said to be similar or closely related to fearfulness. The most well-known example is 'emotionality', which is a trait that has been thoroughly investigated, mostly in rodents. Other 'neighbouring traits' to fearfulness are 'stress-proneness', 'nervousness' and 'timidity'. They all share some, but rather different, facets of fearfulness. More or less, however, they are all constructs that are suggested to describe the individual's general tendency to react to threatening and potentially dangerous situations. This makes this trait highly relevant in several regards, not least from a welfare perspective. Dogs that are generally fearful appraise threats in a range of situations, and it is assumed that they often experience negative emotions such as acute fear and anxiety.

However, the concept of a general tendency to react to threats has been questioned. There are several studies that give evidence for more narrow subtypes of fearfulness. One example is the difference between fearfulness towards social and non-social stimuli; fearfulness towards strangers may not necessarily be associated with fearfulness towards such stimuli as sudden or loud noises, novel objects and thunderstorms. Indications of this come from studies of potential guide dogs that were observed in a range of situations (Goddard and Beilharz, 1984, 1986) and from questionnaire studies (e.g. Hsu and Serpell, 2003). Furthermore, social fearfulness may be divided into different subtypes, for example, fear of unfamiliar dogs and unfamiliar persons (Goodloe and Borchelt, 1998). Also non-social fear tests have yielded results that indicate the existence of several fearfulness traits specific to particular stimuli (King *et al.*, 2003). With the findings of different subtypes of fearfulness follows a questioning of the concept of fearfulness; does one general tendency in dogs to react to threatening and potentially dangerous situations really exist? This is a justifiable question – if there are different tendencies to react fearfully towards different stimuli, why use the concept at all?

Another question regarding general fearfulness is how the dog reacts in threatening situations. Research on other species suggests the existence of different coping strategies; either the individual reacts with an active strategy – fight or flight – or with a passive strategy – 'freezing' or immobility. Such differences in strategies have been reported in dogs. For example, results from several older studies suggest the existence of two types of fearfulness (e.g. Thorne, 1940; Royce, 1955). These studies seem to be inspired by the work of Pavlov, who suggested two types of dogs in this regard: an excitable and an inhibitable type. There is also some support for breed difference in this regard. Scott and Fuller (1965) reported that inhibition is easily elicited in cocker spaniels and Shetland sheepdogs, whereas basenjis are more prone to active avoidance. However, such clear-cut differences in coping strategies have been questioned. It is likely that such differences between individuals exist, but it is probably more a question of tendency than of kind – some individuals might be more prone to inhibition, whereas others might often react with active avoidance. Another factor that interacts with this possible personality trait is the type of stimulus-situation. Some threatening situations might elicit immobility to a higher degree than others, where strategies such as flight are more common.

There is, however, evidence of more general tendencies to react with fear that makes the concept of fearfulness relevant in dogs. Goddard and Beilharz (1984) found one general fearfulness dimension besides several more specific dimensions. Furthermore, fearfulness in social and non-social situations has been found to be correlated, just as fearfulness towards unfamiliar dogs and unfamiliar persons. Results from a study carried out by myself using a Swedish version of the questionnaire CBARQ (developed by Hsu and Serpell, 2003) showed positive correlations between four measures of fearfulness: 'stranger-directed fear', 'non-social fear', 'dog-directed fear/aggression' and 'pain sensitivity'. The correlations were moderate, ranging from 0.20 to 0.32, but indicate that there is a general fearfulness influencing fearful behaviour in different situations.

Aggressiveness

Another highly relevant trait in dogs is 'aggressiveness'. This trait has been suggested for dogs in a number of studies, and might be defined as the dog's general tendency to act threateningly (for example, raised hackles, bare teeth, heightened body posture, raised tail, growling) and aggressively, such as attacking and biting. Two Dutch studies are of interest in this regard (Netto and Planta, 1997; van den Berg et al., 2003). In these studies, a similar test battery was used in order to describe the individual dog's aggressive tendencies. The dogs were tested in a range of subtests, where they were exposed to stimuli situations such as approaching persons, unfamiliar dogs, tug-of-war, handling by the owner, feeding competition and a life-sized doll. The major aim with these studies was to investigate whether the dog's behaviour in the tests reflected the typical aggressive behaviour according to the owners. No direct analyses were conducted in order to find out if there exists a general aggressiveness trait, but the results suggest that there is a general aggressiveness component in dogs – aggressiveness towards both dogs and persons – that is possible to predict in a behavioural test. However, two different types of aggressiveness were found based on the correlations of behavioural reactions (van den Berg et al., 2003). One type was defined as 'threatening' (stiff posture, staring, growling, and pulling of the lip), and one was labelled 'attacking' (barking, baring the teeth, attacking, and, to some degree, snapping). Furthermore, the owner's description of the dog's typical behaviour indicated that aggressiveness towards dogs and persons does not necessarily have to be correlated. This suggests that, besides a possible separation between threat and attack, there are different types of aggressiveness associated with different targets. This is supported by studies where questionnaires have been used. Goodloe and Borchelt (1998) found evidence for three such types of aggressive behaviour: towards family members, towards strangers and towards unfamiliar dogs. Similar types were found by Hsu and Serpell (2003), who used the CBARQ. Their results also suggested an additional type of aggressiveness: towards familiar dogs. Clinical studies suggest types of aggressive behaviour that may be candidates as stable traits. Examples are object-related aggression (defence of food, toy or other object) and territorial aggression (for example, aggression towards persons when the dog is in its own yard).

Thus, the situation is similar to what has been described for fearfulness – the indication of several types of aggressiveness trait raises the question whether a general aggressiveness trait exists. As for fearfulness, correlations between different measures of aggressive behaviour from the Swedish version of the CBARQ indicate the existence of general aggressiveness (correlations between 'stranger-directed aggression', 'owner-directed aggression', 'dog-directed aggression/fear' and 'familiar dog aggression' ranged from 0.15 to 0.32). However, these are only indications. There is a need for studies of aggressive behaviour in a range of situations, such as the test in Netto and Planta (1997), where the issue of generality versus specificity can be raised.

Other personality traits in the dog

Some other candidates of relevant personality traits in the dog are worth mentioning. An often-proposed stable trait in animals is a *general activity level*. Compared to fearfulness and aggressiveness, which might be defined by reactions to a specific class of stimuli, activity is a more unspecific trait. The assumption behind this trait is that 'active behaviour' in one situation, for example measured as the frequency of paw liftings, is correlated with activity in several other situations.

A trait related to activity is *reactivity* or *excitability*. Results suggest a difference between being active in non-stimulating situations and being reactive or excitable when stimulated. In some studies, reactivity has been measured in threatening situations, which suggests that it might be a measure of the dog's fearfulness. But there is some support for a more general tendency to be excited when stimulated. For example, Hsu and Serpell (2003) found a relationship between the dog's excitability in situations such as when the owner returns home, when playing with a member of the household and when being taken on a car trip. Analyses of data from a Swedish version of the same questionnaire that Hsu and Serpell used (the CBARQ) suggest relationships between this type of excitability and other behaviours that indicate a more general reactivity. For example, correlations were found with attachment level to the owner – which the family-related items suggest – aggressiveness towards strangers and aggression and fear towards unfamiliar dogs.

The dog's tendency to be friendly towards unfamiliar persons has been described in several studies. This tendency has been proposed to be a personality trait in the dog, which often is labelled *sociability*. The dimension seems to range from an active and 'friendly' approach to strangers to an attitude of reserve, or hostility, to strangers. The negative side of this trait seems to be related to social fearfulness and aggressiveness, and perhaps it is the positive side – a positive interest and a friendliness towards unfamiliar persons – that motivates a use of a separate sociability trait. A question that seldom has been addressed is whether the dog's sociability towards unfamiliar persons is correlated with the same attitude towards unfamiliar dogs. Results from a questionnaire study made by myself (Svartberg, 2005) showed a correlation of 0.33 between friendly behaviour towards persons and dogs, which suggests a common sociability factor.

A very typical behavioural category in the dog is playing. There are results that support that there are stable differences between dogs in this regard, which suggest an existence of a *playfulness* trait. A trait that describes the dog's tendency to run after a thrown rag, grab it and willingness to play tug-of-war with it has been detected in a behavioural test (Svartberg and Forkman, 2002). This trait is consistent over repeated tests, and correlates with owner reports of their dog's general interest in playing with objects with familiar and unfamiliar persons (Svartberg, 2005; Svartberg *et al.*, 2005). From questionnaire studies, other play-related traits have been suggested. Goodloe and Borchelt (1998) found two playfulness traits that were both object-related and person-directed. One was related to chasing after thrown objects and carrying objects, and one trait could be described as 'vigorous' play, related to growling and shaking while playing with objects, and tug-of-war. Results from Svartberg (2005) suggested one trait that described object-related play with humans and one dog-directed trait with relations to both interest in playing with other dogs and approach of dogs in general. This indicates that playfulness towards persons and dogs might be influenced by separate traits.

Finally, many attempts have been made to measure *trainability* – the dog's general success in training situations. Different studies have used different procedures to measure this trait: from using results from one specific task, such as retrieving, to an assessment of the dog's general performance, including such aspects as reactions to distractions, persistence and cooperativeness towards the trainer. However, the search for the dog's general trainability is so far rather unsuccessful. The results from the major study by Scott and Fuller (1965) are interesting in this regard. They trained dogs of five breeds in several different situations in search for a general 'intelligence'. Because of the aim of their project – understanding the genetic bases of behaviour – they used average breed values when presenting the data. The results were striking. No breed was generally better than the others when taking all tasks into account. For example, basenjis were the best in a manipulation test, but the poorest in a trailing test. The beagle was the best breed in a T-maze test, but was among the poorest breeds in the manipulation test. The only consistent pattern was found for the Shetland sheepdog, which was generally ranked as the least successful breed in the tests. The authors explain this with this breed's (at least in the sample used) relatively large fearfulness. Thus, fear of apparatus and persons inhibited learning performance. Probably, this is characteristic for performance in learning situations. A general trainability factor is difficult to find; performance is influenced by a range of task-specific factors as well as several personality traits. Results suggest that, besides low fearfulness, playfulness and excitability may predict training success. Playfulness probably sets the value of play as a reinforcer, which gives that playful dogs might be easier to reward than less playful dogs.

Excitability, which has been found to predict performance in guard dogs, seems to relate to success in a non-linear way (Martínek *et al.*, 1975). The dogs with the highest level of performance were those that had an intermediate excitability level, whereas dogs that had high or low excitability performed less well. According to the authors, this effect might, at least partly, be due to a higher habituation rate for the moderately excitable dogs. Such factors as fearfulness, playfulness and excitability are probably not equal to a general trainability factor, and might have no correlation with the individual's typical – if there is a typical – learning ability (such as habit formation and memory retention), but might nevertheless predict success in training situations.

Besides these described traits, there are a number of other candidates, although more poorly investigated in the dog. From questionnaire studies, traits such as separation distress, tendency to dominate or act submissively, predatory interest, attachment to owner, tendency to bark and pain sensitivity have been suggested. Further studies might give us more knowledge regarding the stability of these suggested traits in dogs.

Motivational State or Behavioural Reaction?

It should be noticed that the theoretical basis of the suggested traits differ, and this influences the results. The theoretical issue is whether personality has its base in motivational states or in behavioural strategies. For example, if the researcher assumes that there is a fear state in dogs, the experimental design and the way of collecting behavioural data will be influenced by this assumption. Different potential fear-eliciting stimuli will probably be used, and measures of flight distance or latency to contact, or a fearfulness scale, might be used as fearfulness indices. Correlations between the measures will say something about the generality of fearfulness in the sample used. However, an alternative assumption is that personality derives from the individual's typical behavioural strategy, which might be independent of motivational state. If this is a correct assumption, there is a risk that such typical behavioural strategies are missed in a 'motivational state approach'. Assume, for example, that dogs differ in general excitability levels. A fearfulness scale, and even measures of contact latency and flight distance, could miss these differences, as well as the stability of each individual's excitability tendency across situations. On the other hand, if the researcher is focused on typical behavioural tendencies, differences in the individual's tendency to activate different motivational states (for example, how often the dog seems to be angry or fearful) may be missed. One such conflict is between coping theory and the concept of fearfulness. Coping theory deals with the type of behavioural strategy the individual uses when confronted with a threatening situation, whereas the fearfulness concept focuses on the level of assumed fear the individual experiences when threatened. In the list of personality traits in the dog presented in this chapter, traits such fearfulness, aggressiveness, sociability and playfulness are based on motivational states, whereas activity, excitability and trainability are closer to a behavioural strategy perspective. The somewhat contradicting results found for fearfulness and aggressiveness can, at least partly, be explained by the two different perspectives. The differences in focus have been acknowledged rather poorly within animal personality. It is likely that personality might be successfully studied from both directions; however, without taking differences in theoretical standpoints into account, there is a risk of confusion and misguiding results.

Personality at Different Levels

As the mini-review of personality traits above suggests, there seem to be traits at different levels of generality. Personality in both humans and animals seems to be best described hierarchically. At the top of the hierarchy, there are a few general traits that might influence the behaviour in a range of situations. These traits might, in turn, have different facets, or subtypes, that more specifically may describe the dog's typical behaviour. In dogs, there are some suggestions of such general traits. For example, there are results that support two major dimensions in dogs that reflect positive and negative emotions (Sheppard and Mills, 2003). According to this view, positive activation correlates with such behaviour as playfulness, excitability and exploration, whereas negative emotions relate to fearfulness, phobic tendencies and anxiety. These two aspects of personality are analogous to two of the human 'supertraits': extraversion and neuroticism. There are other studies that support that these dimensions might also be found in dogs. Gosling *et al.* (2003) found evidence for these supertraits in dogs and, in addition, two others: agreeableness (lack of aggression, cooperativeness) and openness to experience (trainability, exploration). At an even more general level, a boldness–shyness axis has been proposed. This dimension, which ranges from shy and timid behaviour to an outgoing, active and bold attitude, has been found in several other species, as well as in humans, and might be correlated to both neuroticism and extraversion. As mentioned previously, a boldness dimension that correlates with sociability towards humans, playfulness and fearless behaviour in non-social situations has been suggested for dogs, which may influence behaviour in many aspects of life (Svartberg and Forkman, 2002; Svartberg, 2005).

These suggested 'supertraits' in dogs are interesting in the understanding of dog behaviour. However, when predicting a dog's more specific behaviour, these general traits may be rather powerless. When selecting a potential working dog, or when predicting behavioural problems, the focus should probably be on more narrow traits. Thus, there is a need for knowledge of dog personality on several levels – the aim determines which level is the most relevant.

Another point with the hierarchical view is that specific traits may be more or less correlated. For example, the theory behind extraversion in humans predicts correlations between activity, sociability and dominance. Such networks of correlations, and lack of correlations, might help us to understand how dogs generally act in different situations. An example of a network of correlations is shown in Fig. 11.3. The figure is based on data from a questionnaire study of 697 Swedish dogs from 16 breeds (Svartberg, 2005). Even though this sample may be too specific to use for a generalization for all dogs, the figure shows probable associations between specific traits. For example, (lack of) sociability, aggression and fearfulness towards humans are tightly connected. Aggression and fearfulness directed towards dogs are correlated to a range of other traits, such as non-social fearfulness, fear of and aggression towards unfamiliar persons, general excitability and aggressiveness directed to other dogs in the household. In contrast, owner-directed aggressiveness and tendency to chase (cats, squirrels, etc.) are not related to any other traits in this study.

Development of Personality

One issue regarding dog personality is related to development – change and continuity of personality during the life course. This question is often addressed in

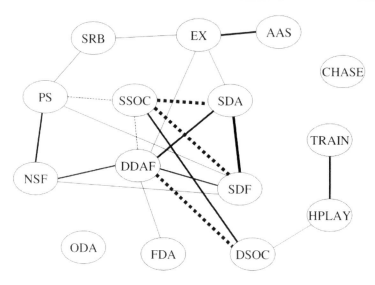

Fig. 11.3. The correlation between different behavioural traits from the study of Svartberg (2005). The thickness of the line indicates the degree of correlation; from the thinnest line that represents a correlation of 0.20 to 0.25, to the thickest that represents correlations above 0.40. Dotted lines indicate a negative relationship (SRB = separation related behaviour; PS = pain sensitivity; NSF = non-social fear; ODA = owner directed aggression; SSOC=sociality towards strangers; DDAF = aggression and fear directed to unfamiliar dogs; FDA = aggression towards other dogs in the family; EX = excitability; AAS = attachment and attention-seeking behaviour; SDA = stranger-directed aggression; SDF = stranger-directed fear; DSOC = sociability towards dogs; HPLAY = playfulness towards humans; TRAIN = trainability; CHASE = interest in chasing).

human personality research. Within the study of animal personality, development of stable dispositions is less often studied. As a starting point for a discussion of what we know about personality development in the dog, three questions regarding personality development in humans formulated by Caspi and Roberts (2001) may be useful: (i) how early in the life course can we identify characteristics unique to individuals that will show continuity and change in personality? (ii) when in the life course is personality fully developed? (iii) what life course factors moderate continuity and change in personality? We are limited to the few longitudinal studies that have been carried out, something that we should have in mind when interpreting the results.

How early can we tell?

Among the relatively few studies where stability of behaviour from young age in dogs has been in focus, fearfulness is the most studied trait. There are some suggestions on very early indications of fearfulness. Royce (1955) used data from a battery of tests carried out within the research programme at the Jackson laboratory (Scott and Fuller, 1965). He found that the average number of vocalizations – whines or yelps – from birth to the third week (observed during weighing) correlated with reactivity score in several situations at the 18th week, as well as with some physiological measures (blood pressure and sinus arrythmia) at 8–9 months of age. Another early sign was a fearfulness score (rated during a handling test) at the fifth week, which correlated with reactivity measures at 1 year of age. A study by Goddard and Beilharz (1986), which was carried out in order to predict fearfulness, activity and trainability in potential guide dogs, suggested that fearfulness was stable from 8 weeks of age. However, the correlation with adult general fearfulness increased with test age: better predictions were made at higher ages.

These results suggest that aspects of the dog's fearfulness may be possible to detect at a very early age. This is supported by a study on wolf pups, where several different tests at 7–9 weeks of age were carried out (Fox, 1972). The results suggested that one general boldness dimension could explain individual differences in a range of situations, including prey-killing, fearless behaviour in different situations and a tendency to dominate other pups. Puppy-boldness, in turn, predicted dominance score at 1 year of age, which suggests that the individual's general boldness, or the tendency to act in a fearless way, is developed at an early age.

The early fearfulness signs, which are promising from a prediction point-of-view, are contrasted by results of other traits. Several studies in this area suggest that behaviour before 8–10 weeks has a low predictive value for the adult's typical behaviour. For example, a study on the predictability of dominance and activity level using a popular test developed by Campbell (1972) suggested no correlation between 7 and 16 weeks of age (Beaudet *et al.*, 1994). Another example is the study of Wilsson and Sundgren (1998), who used data from tests carried out by 630 German shepherds at the age of 8 weeks and at 15–20 months. There were no correlations found between behavioural measures from the early test (vocalization, reaction to a person, reaction to play objects, activity level) and measures from the adult test (however, this result could be explained, at least in part, by differences in methodology between the two tests).

Even though there are relatively few studies in this area, it seems that the predictive power of puppy testing is low. The exception may be fearfulness, which may be able to be predicted at an early age. At what stage it is possible to predict other traits is difficult to assess based on the knowledge so far.

When is personality fully developed?

A common but probably misleading view is that personality develops from birth to a certain age, and then remains stable. In human personality studies, it seems that stability of personality is rather low in childhood, increases in adulthood and reaches a plateau between the ages of 50 and 70 (Roberts and DelVecchio, 2000). However, even though a plateau is reached, personality also continues to change at higher ages. Unfortunately, there are no studies that give us knowledge regarding this issue in dogs. Some studies show high stability of personality traits in dogs over short periods of time (1–2 months) in adult dogs. For example, high rank-order correlations (0.6–0.9) have been shown for sociability towards humans, non-social fear, playfulness towards humans and aggressiveness in adult dogs (Svartberg *et al.*, 2005). However, short-term stability is a basic criterion for personality traits, and does not necessarily tell anything regarding stability over the life span. There are also indications on stability over longer periods of time for these traits (Svartberg, 2005), but how they change in magnitude over the life span, if they change in a predictable way, is not known.

Results from the behavioural test DMA may give us some information regarding the traits that are assessed in the test. Dogs are tested only once, but the differences in scores between dogs of different age might give an indication on how personality changes over time. In order to investigate this question, I used data from one large breed, the German shepherd, where it was possible to group dogs according to test age. Very few dogs are tested at an age above 5 years, but four age categories were possible to compare (1–2 years, $n = 6,113$; 2–3 years, $n = 675$; 3–4 years, $n = 192$; and 4–5 years, $n = 80$). Of the personality traits that are expressed in this test, three were of greater interest: curiosity/fearlessness (expressed as non-social fearlessness in everyday life); sociability; and aggressiveness. As Fig. 11.4 shows, there are differences between age categories for these traits. The same trends were found for both sexes, which indicates that non-social fearfulness, sociability and aggressiveness decrease slightly over the years.

These results, however, are only a rough indication. To really understand how personality changes over the life-span, studies where dogs are followed longitudinally are needed. Hopefully, such approaches will attract researchers in the future.

What factors moderate continuity and change?

The common way of thinking is probably that environment causes change in behaviour during development, whereas the genetic bases of behaviour make behaviour stable. Regarding environmental factors that influence behavioural change, it is likely that a range of factors may contribute to the variation of dog personality. The transition from the kennel to the new home may be an important factor, as well as the family setting in the new environment. The presence of other dogs, number of family members and their age, the non-social environment in which the puppy grows up – countryside or city – as well as how the social environment treats the young dog are probably important moderators of the dog's personality. There are, however, few studies that have focused on the importance of different environmental factors on personality in the growing dog, probably

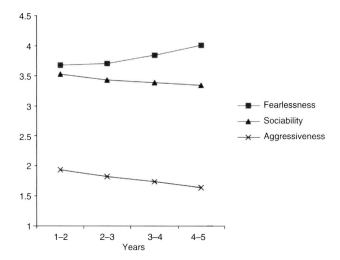

Fig. 11.4. The figure presents the scores for the traits curiosity/fearlessness, sociability and aggressiveness assessed in the DMA test for four categories of dogs based on test age (German shepherds, males and females pooled). Statistically significant decreases in scores with age were found for sociability and aggressiveness, whereas a significant increase in curiosity/fearlessness was detected.

due to the methodological problems of separating different factors from each other. There are some indices from studies where the focus has been on factors associated with aggressive behaviour in dogs. For example, Podberscek and Serpell (1997) found several differences in environmental factors between a high-aggressive group of English cocker spaniels and a low-aggressive group. Owners of low-aggressive dogs were older and more attached to their dogs, whereas high-aggressive dogs were groomed less often and given less exercise. Another study showed that aggressive dogs, compared to dogs that are attacked ('victims'), are more commonly trained by shaking and hitting, whereas a dog owner attitude of believing that dog training should be fun or that training is irrelevant was associated with victim dogs (Roll and Unshelm, 1997). There are some other studies that indicate the importance of owner attitude on dog behaviour. For example, anthropomorphic involvement by the owner has been found to be associated with dominance aggression in the dog, whereas owner anxiety seems to be associated with over-excitement and aggressiveness in the dog (O'Farrell, 1997; Podberscek and Serpell, 1997). Even though these results tell more about how different life situations influence the eliciting of different behaviour, they might also indicate that these factors can be important in the development of a dog's personality.

The view that stability of behaviour during the development is caused by a strong genetic basis probably has its roots in work by early ethologists. Behaviours that were species-specific and difficult to alter with training attracted interest, which contributed knowledge about what has been called fixed action patterns and instincts in a range of species. However, more current research has shown that it is hard to find behaviour in a species like the dog that is totally 'genetic': dog behaviour, in general, is possible to change by environmental factors. However, genetic factors may limit the possibilities, and set boundaries for the development of a behaviour. In some cases, there seem to be very narrow boundaries, which makes the influence by environment rather minor. The degree of heritability of a behaviour (see Chapter 4 in this volume) may give us a hint about how strongly the behaviour is influenced by genetic factors. In a few cases, behavioural traits seem to be strongly influenced by genetic factors. One example comes from an unpublished study by Arvelius (2005), who studied the heritability of herding traits in the border collie using data from a herding test carried out during early training. These results suggested heritability estimates of between 0.40 and 0.55 for several herding traits (for example, the tendency of keeping distance to sheep and anticipation of sheep movements), which suggests that these rather specific traits to a large extent are influenced by genetic factors. Regarding traits that may be considered as broader personality traits in the dog, such high estimates of heritability are rather uncommon. Heritability estimates of 0.2–0.3 are often acknowledged as relevant and rather high. A study of the heritability of the traits found in the DMA is an example of this (Strandberg *et al.*, 2005). The results suggested heritability estimates of 0.23 for curiosity/fearlessness, 0.22 for playfulness and 0.15 for aggressiveness. Such results indicate that there are genetic bases behind personality traits in the dog, even though these should not be over-emphasized. Thus, genetic factors may buffer the impact of environmental influences, and create stability during the development.

This view, however, is not telling the whole story. Genes may create change, and environment may cause stability. One obvious example of gene-created changes in behaviour is sexual maturity. During the transition from juvenile to adult, the genetic expression changes. Some genes that were active during the juvenile period deactivate, and other genes start to produce proteins that influence changes in the dog's typical behaviour. The findings that gene activation is fluctuating, over both the short and long terms, give a contrasting picture compared to the traditional 'stable gene' perspective (changes in gene expression is described in Chapter 4 in this volume). On the other hand, stability in personality may be caused by environmental factors: environmental stability causes stability in behaviour. A dog that changes homes several times during its lifetime may appear inconsistent in its typical behaviour, whereas a dog that lives in a non-changing environment may be assessed as more stable. This perspective is important when applying results from developmental studies. A common goal in scientific studies is to control or wipe out factors that may create variations that are not relevant to the aim of the study. For example, Scott and Fuller (1965) tried to give all dogs a similar environment during their first year of life due to the interest in the genetic basis of social behaviour. This meant, for example, that the puppies never made the typical transition from the kennel to the new owners at about 8–9 weeks of age. In turn, this effect upon behavioural development was never studied, which may make it difficult to apply their results to pet dogs. For example, the breed differences found might be less pronounced when studied in a non-controlled environment.

There is an ongoing interaction between genes and environment, and it is seldom possible to single out one cause for behavioural development. The most striking example of such an interaction is the impact of environment during the socialization period. This period in the dog's life, which is assumed to range from 3 weeks to about 3 months of age, is marked by the large influence environment has on future behaviour. A number of experiments have been done regarding this, which show that several stable characteristics in the adult may be set during this period. Examples are fearfulness towards dogs and humans, reactions to separation, general activity level and trainability (reviewed in Lindsay, 2000; see also Chapter 6 in this volume). If the dog is able to interact with several dogs and humans during this period, the probability of developing high fearfulness in social situations diminishes. If this is delayed to after 3 months of age, the effect of the social interaction will be smaller, especially if the dog has been separated from dogs and humans before this age. Thus, the socialization period is a good example of how genes and environment interact: the genes open up a 'socialization window' during a specified period, but the effect of this is dependent on environmental factors during this period.

The currently emerging research within genetic techniques, which is developing very fast, may soon give us more knowledge of the genetic basis of behaviour – which genes influence which behaviour, and the function of the mechanisms that are in charge during development. With this as a base, we may single out environmental factors – what and when – that are important during the development of personality in dogs. Results from other species are promising. For example, newborn rats that do not get licked and groomed by their mothers grow up timid and highly sensitive to stress (Weaver *et al.*, 2004). The mechanisms behind this effect seem to be an increase in methylation, which suppresses genetic activation. The decrease of genetic activation in a certain gene, a promoter for the glucorticoid receptor gene, causes fewer receptors to be produced, and, in turn, more timid and anxious rats.

Applied Use of the Personality Concept

This chapter has dealt with dog personality mostly from a scientific point of view. However, the concept of personality is also relevant, and has a widespread use, among dog owners, dog trainers and breeders. Besides the traits that have been presented here, there are numerous others. For example, in the breed standards there are large numbers of adjectives describing each breed's typical behaviour; most of these adjectives are similar to personality descriptions used for humans. There are also a number of popular behavioural tests that are used in the prospect of revealing the dog's typical way of acting, from a variety of puppy tests to different tests used to capture the typical behaviour of adult dogs. Evidently, personality in dogs – perhaps more often referred to as mentality, temperament or nature in everyday terms – is relevant for all persons involved in dogs in one way or another. It is important to describe the characteristics in dogs; for the breeder, who cares for a selection of suitable breeding animals and strives for a good breeding result; for the dog trainer, who is interested in selecting potential individuals or matching the dog with adequate training; and for the dog owner, who might want to understand the essence of his or her dog.

While the benefits of using labels and behavioural tests to describe dog personality among dog people are clear, it could be important to highlight some drawbacks. Regarding descriptions of personality in dogs, it seems to be difficult to find a consensus even among scientists, which of course makes it more difficult for laymen to develop a common language regarding behavioural differences in dogs. A characteristic named 'stress-proneness' by one person could be referred to as 'liveliness', 'active temperament', 'high drive' or 'happiness' by others. Alternatively, one label may have different meanings to different persons. A dog with a 'calm' temperament according to a breeder of a pet dog breed may very well differ significantly in its typical activity level compared to another dog with the same description made by a breeder of working dogs. Even such a specific trait label as 'interest in objects' may differ between persons. One might mean the dog's interest in tug-of-war, another the dog's willingness to carry objects in the mouth, and a third the dog's enthusiasm to search for motivationally significant objects (such as toys or dummies). The differences in terminology might be a unsolvable problem, at least as long as the scientists in this field have difficulty reaching consensus. Within the field of science this is partly solved by a routine of defining concepts and terms that otherwise could create misunderstandings (for example, 'sociability' might be defined as 'the dog's interest to interact with other individuals in a positive manner'). My belief is that this strategy would be beneficial for the non-scientific part of society that deals with dogs. The mere knowledge of differences in definition is important to bear in mind in order to avoid confusion; defining terms and labels would be even better. In other words: don't take descriptions of behaviour and personality for granted, try to understand what is meant by them.

When it comes to behavioural tests it is important to bear in mind that the usefulness of results from a test is often very limited. The test situation itself limits the usefulness, something that has been described previously in this chapter. Some traits are possible to detect in a test, whereas others are not. Besides this, behaviour tests arranged by laymen often have several weaknesses.

First, the standardization of the test situations is often poor, so dogs may experience different test stimuli. For example, one dog may be tested in a situation where a person throws a tennis ball away from the dog, a second will experience a ball thrown to the side and a third dog may be put in a situation where the ball is coming towards it. When a test situation has this sort of variation – which could be both larger and smaller compared to this example – it could be difficult to separate the effect of test variation from the effect of variation in personality between dogs.

Second, the way of describing the dog's behaviour in a test is often insufficient. It might be too general to be adequate. For example, a three-step scale with the alternative of 'shows no aggression', 'shows moderate aggression' and 'very aggressive' may miss important differences between individuals (especially if the third alternative is only rarely used). Another issue, which might be more problematic, is that the description of the dog's behaviour or personality can be a mix between intensity and value, or just an evaluation that ranges from 'good behaviour' to 'bad behaviour'. When such a description is made, the person who describes the dog assesses the behaviour against some pre-defined standard. The risk of this is that this standard becomes a 'golden standard' for dogs in general, even though it was meant to be used for a specific object. One example could be that a dog that has carried out a test for working dog aptitude may be regarded as 'unsuitable'. This might give valuable information for the owner regarding working dog use, but it may say nothing about the dog's suitability to be a pet dog, or usefulness in some dog sport. Thus, it is important to look at the standards from which statements are made regarding the value of the dog's personality.

Third, even though it might seem that the test is useful for making predictions of a dog's potential in a certain area – such as working dog, hunting dog, herding dog or pet dog – it is very likely that test results have not been validated against the use for which they are meant to predict. Thus, there is a risk that the test result is a poor predictor for future behaviour and performance. For example, a puppy might be assessed as 'dominant' in a test made by the breeder, but the predictive power for development of dominance-related behaviour in future may be very little, *unless* someone has studied the correlation between test results and future behaviour in an unbiased manner and found positive associations between these two.

In conclusion, personality in dogs is of great relevance for all of us who in some way have dogs in our lives. There are, however, some difficulties in the application of the concept of personality in everyday life. One issue is that we have different ways of describing the typical behaviour of dogs. This may result in confusion and misunderstandings. Better definitions of terms and labels would improve communication. Another issue is the relevance of results from behavioural tests carried out by laymen. Poor standardization, insufficient description methodology and overestimation of the predictive value of the test results are three possible problems. Nevertheless, behavioural tests organized by serious breeders, dog clubs or trainers may give valuable information of a dog's personality. It is, however, important to use the information from these tests with great care, and avoid far-reaching conclusions.

References

Arvelius, P. (2005) Genetisk och etologisk analys av vallningsbeteenden hos border collie. MSc thesis, Department of Animal Breeding and Genetics, Swedish University of Agricultural Sciences.
Beaudet, R., Chalifoux, A. and Dallaire, A. (1994) Predictive value of activity level and behavioral evaluation on future dominance in puppies. *Applied Animal*

Behaviour Science 40, 273–284.

Campbell, W.E. (1972) A behavior test for puppy selection. *Modern Veterinary Practice* 12, 29–33.

Caspi, A. and Roberts, B.W. (2001) Personality development across the life course: the argument for change and continuity. *Psychological Inquiry* 12, 49–66.

Fox, M.W. (1972) Socio-ecological implications of individual differences in wolf litters: a developmental and evolutionary perspective. *Behaviour* 41, 298–313.

Funder, D.C. and Colwin, C.R. (1991) Explorations in behavioural consistency: properties of persons, situations, and behaviors. *Journal of Personality and Social Psychology* 60, 773–794.

Goddard, M.E. and Beilharz, R.G. (1984) A factor analysis of fearfulness in potential guide dogs. *Applied Animal Behaviour Science* 12, 253–265.

Goddard, M.E. and Beilharz, R.G. (1986) Early prediction of adult behaviour in potential guide dogs. *Applied Animal Behaviour Science* 15, 247–260.

Goodloe, L.P. and Borchelt, P.L. (1998) Companion dog temperament traits. *Journal of Applied Animal Welfare Science* 1, 303–338.

Gosling, S.D., Kwan, V.S.Y. and John, O.P. (2003) A dog's got personality: a cross-species comparative approach to evaluating personality judgments. *Journal of Personality and Social Psychology* 85, 1161–1169.

Harding, E.J., Paul, E.S. and Mendl, M. (2004) Animal behavior – cognitive bias and affective state. *Nature* 427, 312–312.

Hsu, Y. and Serpell, J.A. (2003) Development and validation of a questionnaire for measuring behavior and temperament traits in pet dogs. *Journal of the American Veterinary Medical Association* 223, 1293–1300.

Jensen, P., Rushen, J. and Forkman, B. (1995) Behavioural strategies or just individual variation in behaviour? A lack of evidence for active and passive piglets. *Applied Animal Behaviour Science* 43, 135–139.

King, T., Hemsworth, P.H. and Coleman, G.J. (2003) Fear of novel and startling stimuli in domestic dogs. *Applied Animal Behaviour Science* 82, 45–64.

Koolhaas, J.M., Korte, S.M., De Boer, S.F., Van Der Vegt, B.J., Van Reenen, C.G., Hopster, H., De Jong, I.C., Rusi, M.A.W. and Blokhuis, H.J. (1999) Coping styles in animals: current status in behavior and stress-physiology. *Neuroscience and Biobehavioral Reviews* 23, 925–935.

Lindsay, S.R. (2000) *Handbook of Applied Dog Behavior and Training, Vol. 1: Adaptation and Learning.* Iowa State University Press, Ames, Iowa.

Martínek, Z., Lát, J., Sommerová, R. and Hartl, K. (1975) About the possibility of predicting the performance of adult guard dogs from early behaviour – II. *Activitas Nervosa Superior (Praha)* 17, 76–77.

Netto, W.J. and Planta, D.J.U. (1997) Behavioural testing for aggression in the domestic dog. *Applied Animal Behaviour Science* 52, 243–263.

O'Farrell, V. (1997) Owner attitudes and dog behaviour problems. *Applied Animal Behaviour Science* 52, 205–213.

Podberscek, A.L. and Serpell, J.A. (1997) Environmental influences on the expression of aggressive behaviour in English Cocker Spaniels. *Applied Animal Behaviour Science* 52, 215–227.

Roberts, B.W. and DelVecchio, W.F. (2000) The rank-order consistency of personality traits from childhood to old age: a quantitative review of longitudinal studies. *Psychological Bulletin* 126, 3–25.

Roll, A. and Unshelm, J. (1997) Aggressive conflicts amongst dogs and factors affect-

ing them. *Applied Animal Behaviour Science* 52, 229–242.
Royce, J.R. (1955) A factorial study of emotionality in the dog. *Psychological Monograph* 69, 1–27.
Scott, J.P. and Fuller, J.L. (1965) *Genetics and the Social Behavior of the Dog.* The University of Chicago Press, Chicago, Illinois.
Sheppard, G. and Mills, D.S. (2002) The development of a psychometric scale for the evaluation of the emotional predispositions of pet dogs. *International Journal of Comparative Psychology* 15, 201–222.
Strandberg, E., Jacobsson, J. and Saetre, P. (2005) Direct and maternal effects on behaviour in German Shepherd dogs in Sweden. *Livestock Production Science* 93, 33–42.
Svartberg, K. (2002) Shyness-boldness predicts performance in working dogs. *Applied Animal Behaviour Science* 79, 157–174.
Svartberg, K. (2005) A comparison of behaviour in test and in everyday life: evidence of three consistent boldness-related personality traits in dogs. *Applied Animal Behaviour Science* 91, 103–128.
Svartberg, K. (2006) Breed-typical behaviour in dogs – historical remnants or recent constructs? *Applied Animal Behaviour Science* 96, 293–313.
Svartberg, K. and Forkman, B. (2002) Personality traits in the domestic dog (*Canis familaris*). *Applied Animal Behaviour Science* 79, 133–155.
Svartberg, K., Tapper, I., Temrin, H., Radesäter, T. and Thorman, S. (2005) Consistency of personality traits in dogs. *Animal Behaviour* 69, 283–291.
Thorne, F.C. (1940) Approach and withdrawal behavior in dogs. *The Journal of Genetic Psychology* 56, 265–272.
van den Berg, L., Schilder, M.B.H. and Knol, B.W. (2003) Behavior genetics of canine aggression: behavioral phenotyping of golden retrievers by means of an aggression test. *Behaviour Genetics* 33, 469–483.
Weaver, I.C.G., Cervoni, N., Champagne, F.A., D'Alessio, A.C., Sharma, S., Seckl, J.R., Dymov, S., Moshe Szyf, M. and Meaney, M.J. (2004) Epigenetic programming by maternal behaviour. *Nature Neuroscience* 7, 847–854.
Wilsson, E. and Sundgren, P.-E. (1998) Behaviour test for eight-week old puppies – heritabilities of tested behaviour traits and its correspondence to later behaviour. *Applied Animal Behaviour Science* 58, 151–162.

12 Human–Animal Interactions and Social Cognition in Dogs

Ádám Miklósi

Introduction: Dogs in the Human Niche

In biology, or more precisely in ecology, the concept of a 'niche' is a theoretical characterization of a species' environment, and it is understood that strictly speaking no two species can live in the same niche. 'Niche' is also used, as in this chapter, in a somewhat looser formulation to describe the 'psychological and social' environment of humans that is shared to some extent or has some overlaps with dogs. There can be no doubt that dogs and humans have developed a particular relationship over the last 10,000 years. Dogs, along with other animals like sheep, cattle, chickens and cats, have often been 'lumped' together and referred to as 'domesticated' animals. Additionally, there has been a continuous interest in understanding the process of domestication because for many it seemed to be a good case for modelling evolutionary processes (Darwin, 1859). In this research people were looking for general features of changes that could account for all or most cases of domestication, irrespective of the species in question. For example, many noted that there is a reduction in relative brain size when one compares the ancient wild species with its domesticated descendants (Kruska, 1980). Although it is important in its own right, this approach has overlooked the fact that most domestication events took place at different times, and the original causes for domestication were also probably quite different for different species. Even the aims of domestication (selection) might have changed over time. This suggests that it would be perhaps more advantageous to consider each domestication event (species) separately and investigate more closely the actual relationship between humans and the particular species in question.

Although dogs often have been the focus of biological studies of domestication, and they never lost their appeal for the veterinary sciences or researchers interested in the applied aspects of animal behaviour, to many people dogs still represent a kind of 'artificial' animal. Hundreds of ethologists or animal psychologists working and publishing on 'wild and real' animals have possibly shared their lives with one or more dogs at home. Nevertheless, they seemed to have overlooked the possibility that their favourite companion could not only be an object of compassion and love but also a subject for scientific study. This is clearly reflected in the lack of studies on dog behaviour in major journals of the field like *Animal Behaviour*, *Behaviour* or *Ethology* during the last 40 years or so.

Although one might still consider the dog as an 'artificial' animal (note that there is no such definition as what 'artificial' would mean), it is hard to deny that looking closely and with an evolutionary approach at the dog–human relationship we find that the existence of the two species is tied together closely. In any case, the minimal task of an ethologist should be to explain the present situation when about a quarter or a third of the human population in various countries on the Earth share their lives with dogs. Further, we observe that at present dogs come in various 'forms' (breeds) with regard to morphology, genetics and behaviour, and many see this as an analogy to 'adaptive radiation' observed in many successfully expanding species in biological evolution (Parker *et al.*, 2004).

In my view the most important change in the scientific approach has been that researchers have begun to see the dog as a result of an adaptation to a specific human environment. It has been suggested that behavioural changes from the ancestor of the dog to present-day dogs made it possible for dogs to 'invade' or be 'incorporated' into the human environment. In evolutionary terms, where the number of animals in a given species is used to describe its 'success', at present dogs seem to be a very successful species. They not only come in large numbers, but they are also among the few mammalian species that are distributed over all continents (with the exception of Antarctica).

Although in general one can discriminate 'human niche' from 'niches' of other species, looking more closely there are many types of human environments, even in relation to the dog. This might not have been the case at the start of our contact with the dog or some dog-like species, but at present humans can live in various types of relations with dogs. Often this relationship is associated with some dog 'breeds', emphasizing how dogs were able to adjust on an even more 'sophisticated' level to the human environment. For example, some dogs (or dog breeds) are recognized as 'working dogs', e.g. 'sledge dogs' etc. This categorization has some biological validity, but we need to recognize that in this case 'evolution' seems to run at a high speed, and there are other 'niches' that dogs can and possibly will fit into, like working for the blind (or other people with disabilities) or enforcing law (police dogs), living as pets, etc. If one prefers the evolutionary comparison, it is interesting to see that this diversity of human–dog associations is in strong contrast to the relatively uniform type of social structure that characterizes, for example, wolf groups that have been regarded as the best representatives for modelling the life style of the ancestor species. This brings us to the question of what kind of behavioural modifications took place in the dog.

Looking again at the literature on dog behaviour, one might note a further interesting trend. Books written by authorities on canine behaviour most often follow the traditional trend and describe the intraspecific behaviour of various related species such as wolves, coyotes or jackals. In this respect dogs represent an interesting 'case' where human influence/domestication 'mixed up' behaviours that have previously been under 'natural' selection. People advocating this approach are interested in finding out how intraspecific wolf behaviour (how wolves behave and communicate with members of the same species) has been transformed into intraspecific dog behaviour. In contrast, practical books on dog behaviour, most of them written for the 'public', often with the aim to improve 'training skills' of dog owners, place more emphasis on the interspecific (between-species) aspect of dog–human behaviour, that is, how humans should use their behavioural repertoire in order to influence the behaviour of dogs, and they teach us how to 'read' and 'understand' dog behaviour (i.e. Yin, 2004). Although it seems to be the case that interspecific behaviour of dogs (and humans) has some practical importance, this aspect has not received attention from ethologists or animal psychologists. One might even assume that for present-day dogs, interspecific behaviour might be of even greater importance than intraspecific behaviour, because dogs spend more of their lifetime in contact with humans than with members of their own species.

Testing the Bond: Attachment Behaviour in Dogs

Our approach has always been that dogs and humans represent a mixed-species group. However, we have also assumed that these human–dog groups are to some degree more similar to intraspecific groups than simply being composed of two different species feeding together or sharing the costs of an efficient predator warning system. In contrast, we assumed that just like in the case of social animals living in groups, dogs and humans develop close interindividual relationships, which could take various forms such as attachment relationship, dominant–submissive relationship or friendship.

We should admit that for a long time there has been an interest in characterizing the relationship between dogs and humans, or more precisely the relationship between owners and their dog. However, many early investigations took an anthropocentric view, and were interested mainly in how owners perceived their relationship with their dogs or whether this relationship could be considered as a kind of attachment in a psychological sense (Voith, 1985; Serpell, 1996). This work resulted in the development of 'bonding scales', measuring the attitude of people towards their dog or dogs in general. Although this research was very influential, by their nature questionnaire studies cannot tap into the biological basis of the dog–human relationship.

In order to describe the dog–human relationship at a group level one needs studies that are looking specifically at the behaviour of both parts. Luckily, in another line of investigations psychologists have developed observational methods that seem to be able to describe mother–child relationships at the behavioural level. This so-called 'strange situation test' is based on the observation that mildly stressful situations (novel environments) activate the 'attachment system' of the subject, that is, in the case of humans infants, to express an array of behaviours which characterize the actual relationship between them and the mother (Ainsworth et al., 1978).

Observing the parallels between dogs and children in relation to (adult) humans in the group, we decided to apply the 'strange situation test' in order to describe dog–human relationships. The logic of the test is that after having evoked a mild stress by placing the subject into a novel environment, it is observed either in the presence of the owner or a stranger (a person never seen before). The attachment behaviour of the subject will then be evaluated on the basis of comparing the behaviour shown toward the owner and the stranger. There are three critical aspects: (i) the subject should show more affiliative behaviour towards the owner than the stranger (e.g. play); (ii) it should display a preference for contact and proximity towards the owner; and (iii) it should display specific behaviours towards the owner after separation.

A pioneering study in this respect showed that dogs can be tested very easily in a modified version of the 'strange situation test' (Topál et al., 1998), and we have also developed a quantitative measure of their behaviour, which is different from the more qualitative-type of scoring method used by the psychologists to describe mother–infant attachment. This and other subsequent studies (Prato-Previde et al., 2003; Marston et al., 2005) provided evidence that the behaviour of dogs fulfils the above-mentioned criteria for attachment behaviour. At a group level, dogs showed a clear preference for spending more time with the owner both during the test and also during the 'reunion', when the owners returned after having left the dog for a short period of time. During the test dogs preferred to play with the owner. Further, we have found that dogs could be categorized into different 'types' with regard to their attachment behaviour shown towards their owners based on their scores achieved on the three factors, which has been developed following a factor analysis on the behavioural data (see also Chapter 11 in this volume for a similar approach). These factors relate to the effectiveness of evoking 'stress' in the dog ('Anxiety'), describe the dog's behaviour toward the stranger ('Acceptance') and the owner ('Attachment'). For example, by interpreting multivariate statistical results, we have found that in a group of 51 dogs there were two subgroups (12 and 18 dogs) where subjects showed relatively few signs of stress as a result of being tested in a novel environment. Nevertheless, they clearly differed in their relation to the owner and the stranger – dogs in one of the groups seemingly preferred to interact with the stranger and vice versa.

Since then, it has been established that the 'strange situation test' is very robust, that is, dogs tested after 6 months show similar behaviour, and the testing environment has no influence on the outcome (however, it most be a novel place).

Although until now no direct breed comparisons have been done, at least with reference to the Belgian shepherd it is known that individuals can show the same variation in attachment as observed for a population where each dog represents a different breed (Gácsi, 2003).

Communication with Humans

In animal behaviour communication is described as the use of specifically evolved behavioural patterns in order to modify the behaviour of a recipient to the advantage of the signaller. Although in most cases in the animal kingdom such communicative systems are supposed to be under strong genetic influence, learning can still play a part. For example, in the case of the wolf, Ben Ginsburg (1976) has observed that although individuals raised separately from the pack had no problems in displaying species-specific dominant aggressive signals toward humans from the beginning of their life, they seemed to have problems in showing submissive signals later when encountering other wolves. This suggests that whilst behavioural signals of dominance emerge as a result of maturation, the establishment of the submissive signal repertoire and its utilization needs social experience.

With regard to the dog there could be (at least) three potential ways to explain dog–human interspecific communicative behaviour. First, as both humans and dogs are highly social animals and have a range of signals for social interactions, they can simply rely on their common mammalian heritage for communication. However, it is more likely that efficient communication requires members of both species to engage in some sort of learning in order to be able to recognize each other's signals. Finally, one could suppose that behavioural evolution has given rise to specific signals, which are used predominantly in dog–human communication. Note that these three hypotheses are not mutually exclusive, so all three could explain parts of the rich communicative interactions that take part between dog and human.

In the minds of many people, humans are the 'communicative champions' amongst all animals, since they use various ways and means for communication, and have even developed a unique system of communicative behaviour referred to in general as language. From our point of view the question is to what extent dogs have modified their communicative system in order to be able to recognize human communicative signals or in addition to emit signals that are easily (or preferentially) discernible by us.

The role of gestural and facial signals in dog–human communication

Recently, there has been an increased interest in visual communication between humans and dogs. The reason for this might sound somewhat strange. It has been found that apes seem to be relatively deficient in their ability to communicate with their human partners (experimenters) in various tasks involving human visual gestures. Importantly, it has been found that dogs can perform somewhat better in some tasks, so this suggests that in the course of evolution dogs have evolved some new skills that enable them to understand some human signals, and this ability is not shared between us and our nearest living relatives at present (the apes). It should be noted that this is probably not the last word on the comparison between dogs and apes with regard to their performance in human-oriented communicative tasks, but if one compares the behaviour of the dog to the behaviour of the wolf, the picture becomes clearer.

From the dogs' point of view, wolves produce a better comparison. Although it is likely that today's wolves are different from those animals that once gave rise to the dogs, it still seems to be the most interesting comparison. For example, genetically, dogs are most closely related to modern-day wolves. Therefore one could hypothesize that if we minimize the environmental differences (or in other words experience) during the behavioural development of wolves and dogs, the comparison of the two species should show those behavioural differences which are most likely to rely on genetic influence, and characterize those traits as having evolved as a result of dog–human interaction. For this reason, we embarked on a programme providing 'maximum' human socialization experience for wolves and dogs (Fig. 12.1). Altogether 13 socialized wolves were cared for by their human

Fig. 12.1. Hand-raising wolves. The pups enjoyed 24-h contact with the caretaker during their first couple of weeks.

'stepmothers' for 24 h a day until 3–4 months of age. The wolf pups were removed from their mother on days 4–6, before their eyes opened. From this day on they spent most of their day with their carers, who often carried them around and even took them to bed in the evening. After independence, the wolves were trained to walk on the leash, and they were carried around to familiarize them with various places, humans and different objects. They were regular visitors to our university, and travelled by public transport as well as in cars. Each caretaker nursed only one wolf, but arrangements were made for the pups to see each other regularly and play for long periods.

Starting in their third week of life, the wolf pups were regularly observed in a series of behavioural experiments testing for social preferences (dogs versus humans) (Fig. 12.2), neophobia (fear of novelty), reaction to dominance, retrieval of objects and communication with humans, the results of which will be reported later. These tests continued until the 13th week of age, when the animals were taken to a farm where there were other wolves. During the next 6 months the wolves were slowly integrated into that pack, but once or twice a week they were visited by their caretakers, and taken out from the pack for testing and free social interaction (Miklósi *et al.*, 2003).

Response to human-given cues was tested in the two-way object choice task when the subject has to find a hidden piece of food on the basis of human gestural signals (e.g. pointing). The basic idea of this test is to hide a small piece of food in one of two containers. Next the containers are placed on the ground and the subject is offered a choice. However, it is allowed to go for one or the other location; the hider standing between the two containers displays a gesture signalling the place of the hidden reward. Such gestures could have various forms such as pointing with extended arm and hand, turning the head or only moving the eyes to the side (for more details, see Soproni *et al.*, 2002). Significantly better performance exceeding the chance level (being 50%) suggests that in their choice dogs are relying on the human gestures. (Importantly, this seems to be also a good test for investigating the 'Clever Hans' effect in dogs. (Clever Hans was a famous horse that performed various tricks by learning minute body signals of his trainer (Pfungst, 1965).)) By the time of testing the wolves, we had learnt that dogs performed well in similar tests; however, we did not know whether this was the result of some genetic preparedness or extensive experience with humans. Despite all our efforts to provide intensive social experience for the wolves, our lovely subjects performed quite disappointingly in these tests. Wolves seemed to have no problem finding the hidden food if the human stood behind the correct container or indicated the place of the food by touching and manipulating the container for some seconds. This convinced us that our wolves understood the task, and were able to perform well under the given experimental conditions, which is always important if one aims at comparing different animal species in the same task. In contrast to the dogs, however, wolves were quite poor in choosing the correct location of the bait if it was indicated by a pointing gesture, and their performance equalled chance when the human pointed to a container that was approximately 50 cm away from the tip of her fingers. Although this latter type of gesture

Fig. 12.2. When given a choice between a human and a dog, hand-raised and socialized wolf puppies prefer to stay with the dog (a), in contrast to dogs, which choose the human (b).

reduced the performance of dogs, they still seem to be able to perform well over chance level. Further experiments showed that dogs actually do respond to some body parts that protrude the body torso. In line with this, dogs can find the location of hidden food if humans point with the collateral hand (Fig. 12.3). Interestingly, dogs respond also to pointing with the leg, suggesting that they can generalize to some extent when it comes to 'reading' communicative signals.

One obvious explanation for such abilities in dogs is to refer to analogous communicative signals in the species-specific behaviour of dogs. Indeed, the biological basis of understanding the human pointing gesture could be the fact that dogs often assume a particular body position, 'pointing' with their whole body towards some particular location in the environment, and one could easily witness this behaviour in an exaggerated form in the pointers. However, this does not

Fig. 12.3. A typical 'natural' experiment. The human experimenter points at the baited container to a hand-raised and socialized wolf with his caretaker. Usually wolves do not perform well in these tests.

explain the wolf–dog difference because such pointing behaviour is also part of the wolf ethogram.

In the course of the pointing experiments, we noted that it was quite difficult to direct the attention of the wolf to the experimenter performing the communicative action. Based on these observations, we devised an experiment that challenged dogs and wolves in similar ways. Nine socialized wolf pups and nine juvenile dogs of the same age were trained to pull out a piece of meat attached to a rope from a cage. In six training trials, subjects learnt with similar speed to get the food out by pulling the rope either by their mouth or forelegs in the presence of their owner, who was standing behind the animal. No differences were found between the two species in the acquisition of the task, their latency to eat the meat decreased to the same extent as they learnt the behaviour. This training was followed by a 2-min long 'inhibited' trial when the rope was inconspicuously fastened to the wire of the cage so that pulling the rope was ineffective. In this situation dogs looked much earlier and for longer at their owner, whilst wolves looked ahead, in the direction of the food in the cage (Fig. 12.4).

We know from studies on humans that such gazing behaviour is one of the most important aspects of human visual communication, and indeed we found something very similar in one of our earlier studies. Here the experimenter hid a piece of food in the presence of the dog and absence of the owner. However, the food was unreachable for the dog (on a high shelf in the living room). We found

Fig. 12.4. One of the young wolves learns how to pull the string. The caretaker does not interfere with the training but her presence gives the wolf puppy some social support.

that after the owner returned to the room, dogs emitted both visual and acoustic signals aimed at getting the owners' attention and directing it to the location of the food. Taken together, this suggests that dogs are able to modify their communicative behaviour toward humans by using signals that are characteristic of human-specific communication systems (see Miklósi *et al.*, 2004).

The role of gazing in dog–human communication has been underlined in other experiments looking at dogs' ability to recognize human attention. In a communicative interaction it is vital to make certain that the receiver is actually in the state of receiving the signal. The monitoring of attention can be done quite easily by recognizing that the orientation of gaze (and eyes) provides a clear indication for this. Again, apes living in laboratories with humans did not show very sophisticated abilities, but dogs seem to be more inclined to use the gaze and eye cues of humans as behavioural signals for attention. In line with this, dogs preferred to beg from humans who were looking at them, and retrieved objects to the 'attentive-side' of the owners, that is, retrieved objects were presented mostly to the facing owner. Interestingly, dogs seem to be sensitive also to the orientation of the owner when responding to verbal commands. In everyday situations, gazing and command goes hand in hand: we are looking at the dogs whilst emitting a command. It has turned out that dogs do not respond blindly to the commands but also take into account whether it has been accompanied by the appropriate cues signalling the attention of humans (Virányi et al., 2004). If the command was issued while the owner was orienting to some other location in space or to another human present, many dogs did not obey.

Especially the latter is very interesting, because dogs seem to be inclined to obey the command more if there was nobody else present in comparison to the case when the experimenter directed the command to another human in the room. One might suppose that dogs have learnt to use visual cues of attention in order to discriminate situations depending on whether they are in the focus of their owner or other humans. Finally, the dogs' ability to recognize humans' attention is also well known from everyday experience when we are 'competing' with our companions for food that 'lies around', for example, in the kitchen. When modelling the situation in the laboratory, researchers have shown that after forbidding the dog to eat a piece of food (Call et al., 2003) the animals were more ready to disobey the command if they conceived the human as not looking/orienting at them (e.g. she was playing a video game or reading a newspaper).

Communication by sounds: words and barks

For many, the main channel of dog–human communication is acoustic, as humans use verbal signals (mostly parts of language) to influence the behaviour of dogs. This aspect of interspecific communication has a very interesting past in the behavioural sciences, and raised much debate – some have even assumed that (some) dogs might have the ability to understand simple forms of human language, whilst other maintained that dogs do not even respond to the words, but only to some conspicuous acoustic features of verbal signals. Perhaps not accidentally, one of the first truly experimental papers on a dog tested his ability to respond to various verbal commands excluding the possibility for visual cuing by the owner (Warden and Warner, 1928). In this study the owner commanded the dog from an adjacent room, ordering him to sit down, go under the table, retrieve objects, etc. Despite the fact that the situation was very unfamiliar for the dog, he seemed to perform quite well, showing evidence that he made the association between the verbal signal and the behaviour to be performed.

Interestingly, until now there are only few studies investigating to what extent dogs understand human verbal signals. Are they only relying on some general acoustic features of the word (would they mix 'Sit!' with 'Sick!') or do they place more emphasis on the accent, etc.? (See also Chapter 8 in this volume for a further discussion of this aspect.) A further interesting question is whether dogs recognize the structure of command sentences (e.g. Bring the ball!) as being composed of a verb and an object, or the action becomes inseparable from the goal. There are also large individual differences among dogs (even in the same breed) with regard to how well they can be 'trained' to respond to verbal signals. This fact has been demonstrated in a recent case reported in the journal *Science*. German researchers found that a dog, named Rico, could retrieve about 200 different objects after hearing a command (Kaminski *et al.*, 2004). Such a 'vocabulary' closely corresponds to the knowledge of words by some apes trained to understand human visual signals. Further research with Rico showed that he was able to extend his knowledge of object names continuously, and that very little, if any, training was needed for the establishment of a novel word–object association. In one task Rico was provided with a set of objects, some of which were novel for him. First, he was commanded to retrieve 2–3 familiar objects, and then he was given a command incorporating a novel word. In response to the novel command Rico retrieved a novel object in most cases, and moreover this single trial was enough to establish a stable memory about the name of the object, as he was quite successful in retrieving the object after several months lapse, even if he had no further contact with the object in the meantime.

Whilst in this experiment Rico learned the names of the object by listening to commands directed at him, there is now some evidence that dogs can also learn the association between acoustic cues (words) and objects if they witness humans talking about them. The experimental method for showing such ability has been 'imported' from research on parrot communication. Grey parrots seem to be more efficient in learning (and of course using) human words if they are placed in competitive situations, where they have to compete with a human for the attention of the 'teacher' (Pepperberg, 1991). Later in training, of course, not just praise is used, but sometimes also more conceivable forms of reward, such as food. In a somewhat modified experimental situation dogs could witness the conversation of two humans who interacted and exchanged a novel object whilst continuously talking and referring to the object by a novel word. After a short exposure dogs were commanded to retrieve the object. McKinley and Young (2003) found that the dogs showed the same level of performance if they had been trained by conventional (operant) training methods. This observation suggests that dogs have the ability to 'pick out' single cues ('words') from a stream of acoustic signals and understand that these might refer to some parts of the environment.

Independently from what dogs might or might not understand from human linguistic signals, every day dog owners continue to talk to their dogs. True enough, many of us also talk to cars, computers or washing machines (especially if they refuse to 'cooperate'), but probably we would not claim that these machines 'understand' us. In contrast, the majority of owners believe that their dog 'understands' them to a certain extent. In a survey, we asked owners to provide a list of commands and describe the 'content' of conversations they have with their dog. We have found that in most cases owners refer to objects or try to prevent the dog from doing something. Additionally, owners often provide information or ask questions to the dog (Pongrácz *et al.*, 2001). At least according to the experience (or belief) of the owners, in most cases dogs seem to 'understand' the utterance because they show appropriate change in behaviour, either 'always' or given a 'particular situation'. Thus, for example, when opening the car door an owner might say, 'Get into the car!', and after seeing the action performed the owner might deduce that the dog has indeed understood the meaning of the utterance. This would be especially convincing if the dog would hesitate to execute the expected action without the command/request. Although it is more than likely that dogs do not understand the meaning of the request in linguistic terms, it could be that the acoustic utterance contributes to the dog's understanding of the situation.

Recent research has shown that acoustic communication between dogs and humans is not a one-way process. It is well known (and experienced everyday) that dogs are relatively 'vocal' (a dog hater would even say 'noisy') animals in contrast to other mammals. The simple reason is that dogs have a propensity to bark very often in various situations, and barking can also easily be facilitated by other dogs' barking. Barking has long been dismissed as a 'by-product' of domestication, having no obvious function. However, it has turned out that barking might play a role in dog–human or even dog–dog communication.

Playback studies have revealed that humans seem to be proficient in categorizing barks if provided with a list of possible scenarios of recordings, and they were also skilful in describing the 'emotional' content of the bark. This means that if a bark was recorded whilst a dog was 'attacking' a human ('Schutzhund situation' enacted at a dog training school, see Pongrácz et al., 2005), then the same bark was described (only by listening to it) as being 'aggressive'. Importantly, this ability in humans did not depend on previous experience with dogs because we have found no major difference between the performance of dog owners and humans who have never owned a dog.

Comparing the vocal ethogram of wolves and dogs, it turns out that barking is used only in very specific contexts in the case of the former species. According to Schassburger (1993), wolves bark in contexts of warning/defence and protest only, and they usually produce one or a few distinct noisy (atonal) sounds with low dominant frequency (Feddersen-Pettersen, 2000). In contrast, barks are characterized by a wide range of frequencies in dogs and they are often repeated over a long duration. Additionally, dogs 'invented' the harmonic version of the bark, not recorded in the case of wolves. This means that barks are used to express a much wider range of emotions in dogs in comparison to wolves. It is very likely that this change in the vocal behaviour of dogs was the result of interaction with humans (the other 'noisy' terrestrial mammal on Earth!), as humans might have preferred individuals with a similar preference for expressing emotions by vocalizations.

Communication and social cognition

The world-famous cognitive ethologist Donald Griffin used to argue that the study of communication in animals provides us with a window for understanding their cognitive abilities. With respect to dog–wolf comparisons, Frank (1980) has also supposed that there is a major difference in the 'information processing system' of wolves and dogs. He argued that the wolf possesses two distinct information processing systems working separately. The first deals with a narrow band of stimuli associated with behaviours directly coupled to survival. The second system provides a very flexible processing of environmental information, which, among others, offers the possibility for generalization and learning from others by observation (see Chapter 8 in this volume for a further discussion on social learning). In dogs, these two systems seem to have merged, enabling the processing of a wide range of stimuli with regard to most aspects of behaviour and also to show an extended behavioural plasticity in response to environmental challenge. At the level of dog–human interaction, the change in the information processing systems allows for greater trainability in the dog when arbitrary stimuli become associated with different types of behaviour.

The problem of such comparisons of the cognitive processing systems in wolves and dogs is that we know very little about the wild species, especially when it comes to performance in complex problem-solving tasks or communication. For example, there is very little experimental evidence that wolves learn by observation. However, it has turned out that dogs show enhanced performance if they have the chance to witness a human demonstrator (Miklósi *et al.*, 2004). As described above, dogs seem to be able to associate human acoustic signals (words) with actions, but we know virtually nothing about wolves in this respect. Nevertheless, observations indicate that they seem to have a very sophisticated vocal communication system based on single signals and their combinations.

An alternative approach could be to compare dog social cognition with that emerging in human children. Both 'animals' are often raised in a relatively similar social environment, which later also becomes their living space. Moreover, there is a huge amount of information on the social cognitive development of human children that could be used as a reference. The point here is not that dogs and children necessarily share cognitive mechanisms, but dogs could have evolved analogous systems that perform at the same level as seen in children. For example, we have found that the performance of hand-raised dogs can be equated to the level of 1.5–2-year-old children with regard to some versions of the 'pointing task' (see above).

During domestication, changes appear to have occurred at different levels of behavioural and cognitive organization. Dogs might have acquired an ability to generalize to a greater extent than wolves in certain situations – for example, when communicating with humans. This flexibility is perceived by us as dogs being easier to train, and also to influence ('live with') than wolves. However, we should note that this trait is actually analogous to our ability to use an arbitrary set of signals for communication. Therefore one could suppose that the emergence of this ability in dogs might also lead to differences in cognitive processing, allowing for the dogs to cognitively represent more complex or sophisticated understanding of social interactions.

In addition, we have provided evidence that in some cases dog behaviour has changed in the direction of allowing for a more effective interspecific communication with us. Based on the available evidence, this suggests that these 'innovations' are not the result of exposure to the human social environment but may have some genetic origin.

Acknowledgements

This work has been supported by the Hungarian Science Foundation (OTKA T 049615), the Hungarian Academy of Sciences (F01/031) and the EU (FP6-NEST 012787).

References

Ainsworth, M.D.S., Blehar, M.C., Waters, E. and Wall, S. (1978) *Patterns of Attachment: A Psychological Study of the Strange Situation.* Erlbaum, Hillsdale, New Jersey.

Call, J., Brauer, J., Kaminski, J. and Tomasello, M. (2003) Domestic dogs (*Canis familiaris*) are sensitive to attentional state of humans. *Journal of Comparative Psychology* 117, 257–263.

Darwin, C. (1859) *The Origin of Species.* John Murray, London.

Feddersen-Pettersen, D. (2000) Vocalisation of European wolves (*Canis lupus*) and various dog breeds (*Canis l. familiaris*). *Archieves für Tierzucht, Dummerstorf* 43, 387–397.

Frank, H. (1980) Evolution of canine information processing under conditions of natural and artificial selection. *Zeitschrift für Tierpsychologie* 59, 389–399.

Gácsi, M. (2003) Ethological study of dogs' attachment toward humans. Unpublished PhD dissertation, Eötvös University, Budapest.

Ginsburg, B.E. (1976) Evolution of communication patterns in animals. In: Hahn, M.E. and Simmel, E.C. (eds) *Communicative Behavior and Evolution.* Academic Press, New York, pp. 59–79.

Kaminski, J., Call, J. and Fischer, J. (2004) Word learning in a domestic dog: evidence for 'fast mapping'. *Science* 304, 1682–1683.

Kruska, D. (1980) Domestikationsbedingte Hirngrössenanderungen bei Saugetieren. *Zeitschrift für Zoology, Systematik und Evolutionsforschung* 18, 161–195.

Marston, L.C., Bennett, P.C. and Coleman, G.J. (2005) Factors affecting the formation of a canine-human bond. *IWDBA Conference Proceedings* 132–138.

McKinley, S. and Young, R.J. (2003) The efficacy of the model-rival method when compared with operant conditioning for training domestic dog to perform a retrieval task. *Applied Animal Behaviour Science* 81, 357–365.

Miklósi, Á., Kubinyi E., Topál, J., Gácsi, M., Virányi, Zs. and Csányi, V. (2003) A simple reason for a big difference: wolves do not look back at humans but dogs do. *Current Biology* 13, 763–766.

Miklósi, Á., Topál, J. and Csányi, V. (2004) Comparative social cognition: what can dogs teach us? *Animal Behaviour* 67, 995–1004.

Parker, H.G., Kim, L.V., Sutter, N.B., Carlson, S., Lorentzen, T.D., Malek, T.B., Johnson, G.S., DeFrance, H.B., Ostrander, E.A. and Kruglyak, L. (2004) Genetic structure of the purebred domestic dog. *Science* 304, 1160–1164.

Pepperberg, I.M. (1991) Learning to communicate: the effects of social interaction. In: Bateson, P.J.B. and Klopfer, P.H. (eds) *Perspectives in Ethology.* Plenum Press, New York, pp. 119–164.

Pfungst, O. (1965) *Clever Hans.* Holt, Reinhart and Wilson, New York.

Pongrácz, P., Miklósi, Á. and Csányi, V. (2001) Owners' beliefs on the ability of their pet dogs to understand human verabl communication. A case of social understanding. *Current Cognitive Psychology* 20, 87–107.

Pongrácz, P., Miklósi, Á., Molnár, Cs. and Csányi, V. (2005) Human listeners are able to classify dog barks recorded in different situations. *Journal of Comparative Psychology* 119, 136–144.

Prato-Previde, E., Custance, D.M., Spiezio, C. and Sabatini, F. (2003) Is the dog–human relationship an attachment bond? An observational study using Ainsworth's strange situation. *Behaviour* 140, 225–254.

Schassburger, R.M. (1993) Vocal communication in the timber wolf, *Canis lupus*, Linnaeus. *Advances in Ethology* 30. Paul Parey Publication, Berlin.

Serpell, J. (1996) Evidence for association between pet behaviour and owner attachment levels. *Applied Animal Behaviour Science* 47, 49–60.

Soproni, K., Miklósi, Á., Topál, J. and Csányi, V. (2002) Dogs' responsiveness to human pointing gestures. *Journal of Comparative Psychology* 116, 27–34.

Topál, J., Miklósi, Á. and Csányi, V. (1998) Attachment behaviour in the dogs: a new application of the Ainsworth's Strange Situation Test. *Journal of Comparative Psychology* 112, 219–229.

Virányi, Zs., Topál, J., Gácsi, M., Miklósi, Á. and Csányi, V. (2004) Dogs can recognize the focus of attention in humans. *Behavioural Processes* 66, 161–172.

Voith, V.L. (1985) Attachment of people to companion animals. *Veterinary Clinics of North America: Small Animal Practice* 15, 289–295.

Warden, C.J. and Warner, L.H. (1928) The sensory capacities and intelligence of dogs, with a report on the ability of the noted dog 'Fellow' to respond to verbal stimuli. *Quarterly Review of Biology* 3, 1–28.

Yin, S. (2004) *How to Behave so Your Dog Behaves.* TFH Publications.

IV Behavioural Problems of Dogs

Editor's Introduction

In this last part of the book, there are only two chapters, dealing with subjects of great concern to many dog owners throughout the world. When everything works, the relation between a dog and its handler or owner may be joyful and harmonic – when things go wrong, it may become a nightmare. Many dogs are put down every year because they develop behaviour disorders which could have been prevented or cured. In Chapter 13, we get an extensive overview of the most common behaviour disorders among dogs, their background and possible causes, and the available treatment methods.

Even though many behaviour problems of dogs arise as a result of 'psychological causes', as exemplified in Chapter 13, it will always be necessary to exclude the possibility of physical disease as a cause of the behaviour disturbance. Chapter 14 gives an exhaustive review of different disease states which typically cause changes in the behaviour of dogs. Whereas diagnosis and treatment of such conditions must be handled by veterinarians, the knowledge provided in this chapter is of course essential for understanding the full background of the behavioural biology of dogs.

Behavioural Disorders of Dogs 13

Roger A. Mugford

Introduction

The contract between man and dog would seem to be a straightforward one: we provide food, shelter and security, expecting only companionship in return. In reality, the relationships between people and their dogs are complex and demanding for both species. Social systems in canine societies are as varied as they are in humans, so it is not surprising that people and dogs together invent highly individualistic relationships. From an international perspective, we also see large country by country variations in attitudes towards dogs, and also in how they are kept. Less industrialized or agrarian societies emphasize the utility-value of the dog as a guard, herding or hunting animal. In more industrialized societies such as in Western Europe and North America, the dog's role as a companion predominates.

The ease with which domestic dogs adapt and thrive within different home settings is remarkable. One might have thought that the role of companion or pet dog would be more secure today and less psychologically demanding than in former times. However, the writer's experience is that pet dogs are subject to complex, often contradictory demands from their people. It shouldn't surprise us that inconsistencies in human behaviour sometimes produce irreconcilable conflicts amongst our canine companions.

This chapter will explore how things can go wrong in the human–dog interface. The analysis and treatment of behavioural problems in dogs is an active area of innovation and employment by a variety of professionals, be they drawn from a veterinary, bio-behavioural or dog training background. Eccentric or unwanted behaviours are frequently featured in the media illustrating failing owners, whereas in truth they are so often only examples of individual differences between dogs and in their relationships with people.

Human factors in canine behavioural disorders

The domestic dog is a supremely adaptable species; witness the Siberian husky at home as a sledge dog for Inuit peoples being transplanted into the hot climate of Australia or the high-rise apartments of New York and Tokyo. It is remarkable that dogs demonstrate such plasticity of behaviour, to cope and even thrive in such different social and physical settings.

In the 21st century, we are witnessing dramatic changes in the demography and lifestyles of people, and these are worldwide phenomena. Family units are becoming smaller and in many cases people are living alone rather than within extended families. With smaller family size comes the likelihood that the relationship with a pet dog becomes more focused, more intense and even a substitute for relationships with other people. In the UK, 28% of households are occupied by only a single person, and we see similar social trends towards solitary living in most Western countries. Dogs living in small homes are liable to be deserted or left alone by their single owner for long periods of time if he or she is working, vacationing, etc. It shouldn't surprise us that in this supremely social species, many dogs are intolerant of isolation and so become distressed when left alone.

Another of the conflicts which people impose upon dogs is inconsistencies in day to day provision of physical contact, love, gestures, recognition and management rules. All owners are guilty of this, when on the one hand they stroke or otherwise attend to their pet, say for jumping up in greeting; but the next reprimand it for doing so. *All* interactions, both positive and negative, can have potent effects upon dogs' behaviour. Being chased for 'stealing' an item is rewarded sometimes, reprimanded on others. Inadvertent reinforcement of such commonly performed undesired behaviours as barking, jumping up and even aggression are often seen by canine behaviour therapists.

With increased urbanization and affluence have come fashions for dog-unfriendly lifestyles in the 'perfect home': clean, ergonomic and efficient. Contemporary trends in interior design can have a distinctly adverse effect upon the quality of canine life. Overly clean, smooth floors leave dogs unable to walk comfortably upon say, wood laminate or polished tile surfaces. The behavioural consequence of living upon such slippery floors may be to increase the impact of noise in the home as well as to motivate the dog to find refuge on furniture. This may bring unintended conflict with the owner who prefers his dog to live on the floor. Another example of adverse impact by humans upon their pet dogs is increasing exposure to mechanical, electronic and magnetic devices *within* the home. The resulting sound pollution and electromagnetic changes have not received systematic study in relation to animal behaviour. However, at an anecdotal level, the writer has seen many cases where dogs exhibit signs of distress within the home due to the operation of, say, a microwave oven. In such cases, the symptoms disappear when the dog is outdoors or the microwave removed. For dogs with high hearing sensitivity (e.g. some border collies), life in the home can be an unhappy experience. Household cleaning products with unpleasant or powerful scents are obnoxious to many dogs, yet the manufacturers of such products are unlikely to have run preference/aversion tests upon the domestic animals which will be exposed to these chemicals in the homes of consumers.

Finally, there is the issue of health and hygiene. An over-concern with cleanliness in the home and exaggeration of zoonotic risks from contact with pets discourages normal, tactile contact between people and their dogs. Cultural taboos such as that a dog not be allowed in the bedroom, let alone on the bed, are spuriously justified on the grounds of hygiene as well as imagined impact upon social rank. But to be separated at night removes one-third of the potential time that might otherwise be available for man–dog interactions. Such owners need to be reminded that dogs are, after all, animals with rather straightforward biological and environmental needs, including needs for active, quality interactions by day, and to sleep with companions at night.

Man–dog communication

Domestic dogs become extremely skilful observers of both vocal and non-verbal communication by their humans. They are responsive to both the emotional tone of voice, and its content. Facial and body gestures by owners greatly influence a well socialized dog, as also do more subtle chemical stimuli arising from sweat and other body secretions. These are learned skills which will be less well developed in dogs raised in institutional environments such as kennels, laboratories or denied significant contact with humans as puppies. The tiniest of gestures or change of mood in a human can communicate important messages to a dog, such as that the owner is about to leave on a vacation, embark on a walk with the dog or is about to arrive at journey's end whilst driving.

The remarkable observational abilities of dogs are probably not matched by our ability to read and to understand canine body language. The task is made more difficult by the great morphological differences between various breeds of dog – witness extremes of head and body shapes between breeds like the bulldog, border collie or borzoi.

The schematic outline of body and facial gestures in a 'normal' oligocephalic dog is shown in Fig. 13.1. Note that all parts of the body are involved in these communicatory gestures, from position of ears, eye, musculature and pigmentation of the muzzle, exposure of the teeth, the hair, standing position and tail. Many gestures are likely to be misunderstood by even experienced owners. For instance, a roll over and a belly display is often depicted as representing 'submission', fear or deference towards a more dominant individual. However, the same gestures can become ritualized by learning processes as to be part of a playful, even sexual display. Gestures such as licking of the nose or raising of one paw may

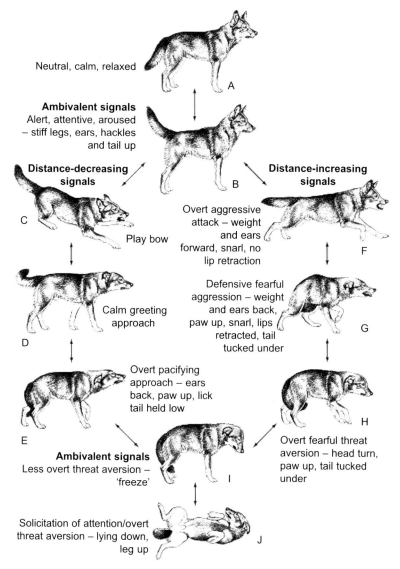

Fig. 13.1. The body language of dogs, related to different social attitudes. Reproduced with kind permission from BSAVA.

have enormous significance for particular dogs, but less in others. These are part of a so-called appeasement complex of gestures, designed to invite approach and friendship in potentially threatening encounters.

A detailed description of canine signals and their significance in social behaviours is provided by Shepherd (2002). These are of especial importance to owners when discussing aggressive or threat behaviours by dogs. Most dog bites are

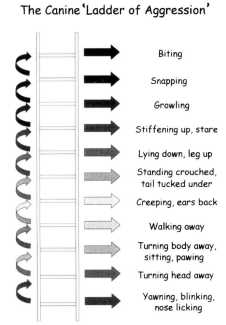

Fig. 13.2. The 'ladder of aggression' in dogs, outlining the different escalation phases a dog will typically go through in connection with aggressive behaviour. Reproduced with kind permission from Kendal Shepherd.

evoked or precipitated by inappropriate actions of humans. The important skill needed by people interacting with dogs is to recognize the interaction between signals representing threat-reducing 'appeasement' gestures versus those signalling defensive, even offensive, threat. Shepherd's 'ladder of aggression' in Fig. 13.2 shows the fluidity or interconnectedness of dogs' reactions which are alternatively threatened or threaten.

Consequences of behavioural disorders

Numerous statistics attest to the life-threatening penalty for any companion animal that presents a behavioural problem to its owner. In this hygiene-conscious age, a dog that eliminates in the home is liable first to receive unremitting punishment, then to be excluded to the outdoors, then finally to be euthanased. Aggression or injury to the owner by the dog is similarly a strong predisposing factor to being euthanased. Estimates of the proportion of young and juvenile dogs euthanased vary between countries, but most studies suggest that more than half of such requests for euthanasia are because they engage in inconvenient or dangerous behaviours (Stead, 1982; Sigler, 1991).

Another, obvious consequence of misbehaving is that the dog is rehomed. The perfect lifestyle in a perfect home is not compatible with, say, a labrador that sheds hairs, occasionally defecates, often smells, and chews the furniture when left alone. And yet, any informed observer would find nothing 'wrong' with the dog; problems arise because of mistaken human expectations and conflicts of lifestyle.

Nowhere on the planet is entirely free from automobiles and cars pose the greatest tangible threat to the freedom and enjoyment of dogs. Because the risk of dogs being killed by cars is so obvious, they must be restrained on leashes and escorted from here to there by humans. Tolerance of being leashed is hardly a trait that featured strongly in the long-term domestication of the dog and it may create a host of frustrations for dogs, at the very least preventing their exchanging normal body signals with other dogs.

Consequently, dogs that are aggressive or intolerant of other dogs represent over 50% of aggression cases referred to our Animal Behaviour Centre. Training such dogs to recover their normal social skills is a relatively straightforward matter, using equipment such as long lines or extending leashes, headcollars, harnesses and well-timed reinforcement of tolerant signals amongst trained, non-threatening stooge dogs. To own a dog which is aggressive towards other dogs can produce a dramatic decline in quality of life or enjoyment of the human–animal relationship. Owners become social outcasts whilst walking their dog, often going to deserted spots early in the morning or late at night. The pleasures of meeting and being amongst other dogs and their owners are taken away by having a pet liable to fight.

Owners of dogs with unwanted behaviours are often made to feel failures by their contemporaries and guilty that it was somehow 'their fault'. As we shall see later in this chapter, the underlying causes of behavioural problems are many and varied, and are not necessarily *directly* attributable to their owner's behaviour. But, of course, sometimes they are!

Finally, there is the matter of legislation, which in most developed countries makes owners liable for dangerous behaviours in their pet. Dogs which are dangerously out of control and might injure somebody can bring prosecutions for criminal behaviour upon their owners in the UK, as in many other countries. Most countries have introduced such legislation, though the boundaries of what constitutes danger or aggressive behaviour is often subjective and dealt with by the courts on a case by case basis. Confusingly, in the USA such legislation is not enacted at a federal level nor even on a state by state basis, rather by local county, city or town ordinances.

Some countries have legislation based upon particular breeds, such as dogs above a given height or bodyweight, alternatively of breeds such as American pit bull terriers, which were originally bred for fighting. Such breed-specific legislation usually has no rational statistical basis for enactment and produces real fear and unhappiness when, for instance, a breed such as the Staffordshire bull terrier is confused with the superficially similar pit bull terrier. These matters frequently appear before the courts in all Western countries, where expensive and emotionally charged confrontations are enacted with the possibility of a death sentence. It is a measure of the love and determination of dog owners that so many owners strongly defend their pet in the face of such seemingly dog-unfriendly legislation.

Examples of Behavioural Problems

A popular adage amongst dog folk is that one man's behavioural nightmare is another's perfect pet. A border collie that chases and attacks joggers will probably also be a good herding dog, good at flyball and heelwork to music. A dog that barks at night may be a useful alert to intruders, alternatively an irritation to insomniacs. So much of what constitutes a behavioural disorder or problem is in the eye of the beholder and there can be no operational definition except by reference to the wants or expectations of the owner. To treat the problem, however, the wants and expectations of the dog have also to be taken into consideration.

There has been a tendency to medicalize the subject of canine behaviour therapy (Overall, 1997), to apply precise-sounding diagnostic terms such as 'hyper-attachment syndrome', 'obsessive compulsive disorder' (OCD) or 'dancing doberman syndrome', etc. The use of behavioural 'labels' or 'diagnoses' such as these is a subject of fervent debate amongst veterinarians and animal behaviour therapists: does such a label disguise our ignorance of underlying causative mechanisms? This writer's view is that the boundaries between abnormal and normal in canine behaviour are so blurred as to make the debate pointless. Better to fully describe the behaviour and its context rather than to prematurely label it as a syndrome. There are, of course, many clinical and physiological factors which contribute to changes in the behavioural or emotional state of dogs. Of these, pain is the most important factor (see later), as also are a number of endocrine and metabolic factors.

The types of behaviours about which owners complain are many and various, but referrals to behaviour specialists such as the writer are overwhelmingly concerned with aggression-related problems. Separation problems are essentially a reflection of a normal attachment process, where a dog misses its human companions and in distress might howl, lose accustomed bowel/bladder control or engage in destructive chewing. Elimination problems or failures to learn to urinate/defecate in approved places at approved times are also a common source of complaint by dog owners, especially where an owner's lifestyle requires that a dog does not have easy access to a garden or other outdoor area. Finally, a large number of behavioural problems are triggered by fear or anxiety. These may be of a generalized nature, where the dog is afraid of many and most everyday situations; alternatively, there may be an acute and overwhelming fear of one or a few stimuli. The latter are known as phobias, and the most common canine phobia is towards loud noises such as from gunshot or fireworks.

Methods for treatment of bang phobias in dogs are now well developed by a combination of desensitization tapes, some means of physically attenuating the dog's hearing, clicker training and possibly short-term use of anxiolytic drugs, including homeopathic remedies.

The skilled behavioural practitioner requires detailed information about the problematic behaviour and a detailed questionnaire may usefully be employed, e.g. BSAVA (2002). This defines the nature of the behaviour which the complaint is about, where it occurs, when it occurs and who in the family is most affected. Answers to these questions will probably then reveal why the behaviour occurs, and from that will come pragmatic strategies for behaviour modification.

Strategies for Treatment

The practice of animal behaviour therapy is both an art and a science. It has become a popular area of employment, with a caché rather like that of some affluent people hiring a personal fitness trainer. Many of those who offer behavioural services have no formal biological or clinical training, yet are free in describing themselves as 'behaviourists', 'behaviour consultants', 'dog listeners', etc. There is conflict between veterinary and scientifically trained practitioners versus those who come to the field as a vocation or as a pleasant, profitable route to self employment.

Regulation of the field has been slow in coming in all countries, and registration of qualified personnel is voluntary everywhere. Distinguishing between a competent and qualified individual versus someone unqualified or with spurious qualifications is difficult for dog owners and they should seek a referral or recommendation from the pet's usual veterinarian. Best qualified of all will be a veterinarian with an interest or additional training in behavioural management.

The starting point for every behavioural consultation is the taking of a comprehensive history that reflects both the events that have affected the dog, also the owner, his or her home setting, lifestyle, etc. Is there a unanimous view within the family about the dog's behaviour or is the pet a focus for conflict and argument between family members?

The setting chosen for preliminary interview needs to be one where the owner is relaxed and not subject to interruptions from phone calls, etc. This *may* usefully be at the owner's home, alternatively in an informal consultation room where the dog can be observed unleashed. Ideally, this will not be a place where the dog has previously had adverse experiences, i.e. probably not in a 'normal' veterinary surgery.

Home visits have the advantage of the therapist being able to witness the dog's home environment and usual behaviours. Critical interactions between dog and family probably only occur in the home. Disadvantages are that the therapist cannot call upon help from colleagues, won't have access to specialized equipment and has to commit time and expense to travelling. However, clever use of video film, comprehensive questionnaires and an informal, varied environment (the author works from a farm) makes the 'clinic' consultation a good first step that can be followed by a home visit if further information or assistance is needed.

To be effective, behaviour therapists must be both skilful in eliciting information, have good listening skills and the ability to resolve issues and arguments within families. The behaviour counsellor is not there to pass judgement, rather to address a problem of enormous importance to the dog owner. He or she may also have an important impact upon the dog's welfare and enjoyment of life and so may need to be explicit about consequences for a pet if improvements are not seen. Many owners feel self-conscious that they have had to seek professional help at all about their pet's behaviour, and are very worried that there might be an adverse outcome (e.g. that there may be a recommendation that the dog be rehomed, or even euthanased).

In the USA (where distances to be travelled are much greater than in most European countries) a telephone interview or consultation is sometimes provided by, say, university veterinary undergraduates. Alternatively dogs are sent away from the owner to be retrained at a specialized kennel facility. The writer is dubious about either approach, given that canine behavioural problems are mostly failures of relationships between dogs and people. The telephone will probably not reveal the relevant social and behavioural dynamics, and there should be total involvement of the owner in all aspects of behaviour modification.

The time taken to fully understand the ontogeny and consequences of an unwanted behaviour can require a surprisingly small investment of time by the therapist. In the writer's practice, most cases require no more than a 2-h consultation, with follow-up by phone, e-mail and possibly a second visit. At my Animal Behaviour Centre, we expect an 80–90% problem resolution rate as defined by the criterion of owners being satisfied with the advice given, and comparable rates are described by other professionals in the field. Such statistics imply that there will be failures, i.e. some dogs will continue to be dangerous or distressed. They may present chronic fear of bangs from gunfire and fireworks, which may be impossible to predict and to retrain acceptable responses. Other dogs present behaviours at such a high level of risk that euthanasia must be the right option, given the behaviour practitioner's responsibilities for safety of the wider public.

Canine behaviour therapy is a dynamic and youthful discipline, given its 'birth' by Tuber *et al.*'s (1974) 'modest proposal'. There are now regular international scientific and veterinary meetings on the subject, many internet websites, journal and book publications. The big challenge for the future lies in integrating this young discipline with everyday education and practice of veterinarians and to 'professionalize' the quality, conduct and content of services provided by behaviour therapists who do not have veterinary qualifications.

Methods

Most professions have agreed, tried and tested approaches, for example to cardiac surgery, dentistry or even car repair. However, the practice of animal behaviour therapy is remarkable for the diversity of methods and their underlying theoretical assumptions. In the world of behaviour therapy, it is a rarity for there to have been systematic study or comparison of the effectiveness of procedure X versus Y. Rather, each case tends to be treated as its own control, and the benchmark of effectiveness is usually reduction in frequency or intensity of some behaviour compared to pre-treatment levels. Most practitioners acknowledge that there is a need for more objective, systematic studies of the effectiveness of methodologies, but understand the difficulties given the unique and interactive nature of animal behavioural problems. I will here outline some of the key methodological approaches that are employed in canine behaviour modification.

Instrumental learning

Instrumental learning is one aspect of Skinnerian theory (named after B.F. Skinner) and is also described as operant conditioning. Instrumental learning occurs whenever a dog's behaviour affects its environment in such a way as to have obvious and meaningful consequences from the dog's point of view. If such consequences are favourable, the behaviour which produced them will be rewarded, reinforced and be more likely to be repeated in the future. If, on the other hand, a behaviour has an immediately adverse outcome, it will be less likely to be repeated in similar circumstances and is said to have been punished. The definition of a reward, therefore, is a favourable outcome which results in an increase in frequency of the preceding behaviour. The definition of a punishment is an adverse outcome with a subsequent decrease in frequency of the causative behaviour.

The ways in which a dog's behaviour can be moulded by outcomes are summarized in Fig. 13.3. It can be seen that the denial of a reward can be punishing (for example, withholding a food reward if a dog jumps up instead of sits), whereas cessation of an aversive can be rewarding (as when an owner ceases to be angry once the 'sit' command is obeyed).

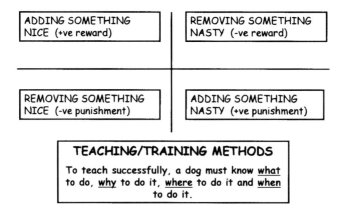

Fig. 13.3. A summary of the main contents of training methods used to alleviate behavioural problems – see the text for details.

The owner of a dog is merely one feature of a dog's environment which can affect and be affected by the way their dog behaves, a fact which frequently needs to be emphasized during counselling. Dogs learn instrumentally all the time, not just when a human decrees that they should during an 'official' training session. Educating dogs to perform desirable behaviours, as opposed to undesirable ones, involves the skilful management of consequences in both training and behaviour modification. Rather than merely punishing undesirable behaviour, owners need to be encouraged to be patient and observant; to notice and to reward all spontaneously performed desirable behaviours.

For either rewards or punishments to be effective, they must be directly linked in the dog's mind as being a consequence of their behaviour. Unpleasant or aversive treatments of many forms are frequently applied by dog owners in attempts to punish behaviour, but they have no impact upon subsequent performance. Paradoxically, the undesired behaviour may actually increase rather than decrease in frequency. In such cases, the aversive applied may inadvertently reward rather than punish the dog and it can have a damaging effect upon the relationship between dog and owner. For instance, hitting a dog can fairly be described as 'retribution', but it is unlikely to guide the dog towards the goal of better behaviour. Attention, even with the pain from a slap of hand, is unlikely to be seen as rewarding.

Fundamental to instrumental learning theory is the notion of schedules of reinforcement, be they continuous, intermittent, partial and so on. A behaviour which is only occasionally rewarded is less likely to cease or be extinguished. On the other hand, continuously reinforced behaviours are rapidly acquired but liable to quickly stop when rewards stop.

Chaining or linking between different responses is another concept drawn from operant conditioning and which is a useful notion in dog training. Complex sequences of behaviour can be brought together for performance of integrated tasks, such as when training a dog that assists people with disabilities to, say, empty a dishwasher or remove items from supermarket shelves.

Importation of learning theory into dog training has had many beneficial effects, especially clicker training. Clicker training was first employed for the better training and precise targeting of behaviours in dolphins. Provision of a distinct click sound associated with the positive experience of being given a food reward enabled precise movements, gestures or behaviours to be chosen and reinforced by the trainer, and consolidated into the animal's repertoire.

Clicker training in the world of dogs has become a major activity, both amongst performance dogs used in film, in specialist tasks such as explosives detection and amongst pet owners who just want their dog to perform tricks. It can also be used to 'shape' desirable behavioural patterns in dogs undergoing therapy. For instance, tolerant, non-threatening signals in a dog that is usually intolerant of other dogs.

Obedience training
Obedience training is a practical application of instrumental learning, where management or control of a dog's behaviour by voice and other signals is a key

part of the behaviour therapist's tool kit. A dog's only problem may be that it chases and kills farm livestock, but one approach to managing such a dog is to ensure that it has a reliable recall response. Similarly, a dog which attacks people as they enter his owner's home may be trained to lie 'down' and stay in sight but out of reach of the visitor. By doing so, one creates a window of opportunity for the guest to make positive gestures towards the dog, such as to offer titbits.

The world of obedience training is *very* idiosyncratic, but has improved considerably from 19th and 20th century ideas that dogs should be dominated and *made* to obey. Writers and trainers from the last century such as Konrad Most, Koeler or Barbara Woodhouse harshly punished clients' dogs on choke chains, but such physical abuse is now generally considered to be distinctly out of fashion. Instead, positive, reward-based methods are more favoured. However, this writer is of the opinion that the move towards *wholly* reward-based methodologies may have gone too far, and something in between the two extremes is better. But if an inappropriate behaviour is to be punished, what is an acceptable and what an unacceptable form of punishment?

Conditioning, both positive and negative, produces distinctive emotional responses in dogs which can have profound and long-lasting effects upon behaviour. A potent punishment for dogs is to remove affection or attention of the owner. The reverse strategy of increased attention is a potent reinforcer of behaviour. Finding the optimum balance between the two provides the best avenue for behaviour modification.

Endocrine management

Castration and ovario-hysterectomy of dogs and bitches is the most widely performed veterinary procedure in the UK and North America. However, it is relatively rarely offered by the veterinary community in Scandinavian countries. Considering the popularity of spay-neutering dogs, few studies have been conducted to explore their effects upon behaviour, good or bad.

Comparisons with other farm animals, such as castrated stallions and bulls, predict a worthwhile reduction in aggressiveness of dogs which have been neutered. There is some anecdotal support for this view, with reduced fighting of other male dogs. However, many bitches also fight and many male dogs continue to present aggressive signs after they have been castrated.

An obvious benefit to neutering bitches is that their liability to subsequent life-threatening diseases such as pyometra (an infection of the uterus) and mammary gland tumours are markedly reduced if the procedure is conducted before the first season. Spaying of bitches is also a worthwhile management procedure where unwanted behaviours occur at particular stages of the bitch's normal endocrine cycle. Pseudopregnancy (elevated progesterone and prolactin titres) can in some bitches produce distinctive, unwanted behavioural changes, such as hoarding of soft toys and guarding of their beds. The recommended approach then is to interrupt the pseudopregnant state with drugs such as cabergoline (Galastop), and to spay 2–3 months after the cessation of pseudopregnant signs. However, to spay a bitch during her pseudopregnant state is not recommended and can cause unwanted behavioural signs to persist (Lindley et al.).

Pain management

Pain has a major influence upon mood or temperament of dogs, just as it does on people. Lindley (1997) estimates that one-third of dogs referred to her behaviour practice exhibit local or diffuse signs of pain. Clinical causes of such pain are many and varied: hip dysplasia, arthritis, intervertebral disc disease, chronic otitis, untreated dental or gum disease and so on.

The prevalence of such behaviour–clinical interactions in dogs presenting unwanted behaviours makes a clinical or veterinary training of the behaviour therapist desirable, and in many cases essential if the best is to be done for the patient. Too many dogs are misdiagnosed as 'dominant' or 'wilful' because of a failure to make a proper clinical diagnosis (see Hedhammer and Hultin-Jäderlund, Chapter 14 in this volume).

Elimination of pain is an ideal which may not always be achievable. Introduction of NSAIDs have greatly improved prospects for pain reduction. There are also alternative therapies such as acupuncture, magnetic therapy, physiotherapy, Tellington-Touch and so on; all approaches which address the underlying cause to disturbed behaviour rather than just focus upon its symptoms.

There may also be indirect but nevertheless significant effects of illness upon the pet–owner relationship. Nursing or care for the injured or otherwise sick animal alters the dynamic between the two: excessive, mistimed sympathy may have profound, unintended consequences.

Dietary management

Interaction between diet and behaviour is a controversial and sometimes eccentric area of opinionated dogma, without the benefit of scientific investigation. Nevertheless, there is little doubt that some dogs dramatically change their temperament or mood when shifted from one diet to another. In the author's practice, it is usual to shift a dog from its commercial or prepared rations to a simple two-ingredient 'elimination diet' such as rice and turkey or lamb and potato, for a test period of say, 1 week. This approach is very like that employed in dermatology and other veterinary specialties where diet or dietary allergies are believed to be an important factor.

Responses of dogs to particular ingredients of either a home prepared or commercial diet tend to be unique to that individual. Certain breeds (e.g. the golden retriever) seem to be more prone to these dietary upsets than others, perhaps reflecting their genetically inbred status and lowered immune competence. It is interesting that golden retrievers are also prone to a host of other immune-mediated diseases, including lymphosarcoma and widespread, chronic eczematous rashes.

The feeding of raw food to dogs (BARF) is becoming increasingly popular worldwide, allegedly improving both the physical and behavioural state of the animals (Billinghurst, nd). However, it is not clear whether the claimed benefits of raw diets are because they are indeed raw, or that various potentially harmful additives, colorants, non-digestible gelling agents, etc. are absent.

Where diet is the factor determining a dog's mood, temperament or behaviour, the effects of changing the diet can be remarkably rapid. Improvements may occur within 24 hours and further improvements continue over time. However, such dogs may still present a liability, evidenced by one golden retriever which was referred to the author's practice. An elimination diet of rice–mutton changed this dog from being a moody, depressed and intermittently aggressive animal into a normal, relaxed and friendly golden retriever. However, some months after the dietary change it gained access to the household garbage bin and dramatically reverted to being a dangerous dog that savagely attacked a 16-year-old babysitter.

My overall conclusion is that dog owners are well advised to read ingredient panels of commercial dog foods. From time to time, they should change brands or feeding schemes and observe the behavioural outcomes. Only a minority of dogs may show beneficial responses to dietary changes, but for those individuals it is as important as is the removal of tartrazine and other colorants from the diets of reactive children.

Physical aids/equipment

Mention has already been made that evolution did not necessarily equip dogs to be made captive on a collar and lead. As one would expect, there has been a host of devices to aid the bodily management of dogs, particularly to stop them pulling.

Choke chains exert substantial compressive forces upon the neck as a whole, with consequent damage to soft tissues of the throat and, if used roughly, to the integrity of the cervical spine (neck). Misplacement at the junction of the skull to the spine (the atlanto-axial joint) is a common sign of whiplash injuries to dogs induced by over-boisterous handling.

Collars with inward-facing spikes were devised in Germany in the last century, and have become popular in North America as a means of preventing dogs pulling. The danger of injury to the dog's throat is obvious and the author has encountered one Weimaraner bitch whose trachea was actually punctured by such a device. These spike collars unfortunately do 'work' (i.e. they discourage pulling), but for the wrong reasons!

Designers of better dog equipment often find inspiration in handling techniques and equipment devised for, say, horses or even for the management of recalcitrant wild animals in zoos. Headcollars were one such innovation, introduced as a standard item for pulling dogs approximately 20 years ago. Numerous designs are now available, some acting as a simple steering device (e.g. the Halti and Gentle Leader), others by exerting compressive forces upon the throat or dorsal neck (e.g. Canny Collar and Nu-trix).

Overall, headcollars are very effective aids to reducing pulling by giving a greater degree of physical control over the direction and movement of the dog's body. Simply put, the body follows where the head is led.

The disadvantage of all headcollars is that they restrict free movement and exchange of facial–body signals by the dog. Members of the public may mistakenly believe that a dog wearing a headcollar is muzzled and so might be dangerous. These problems are overcome by using harnesses, which exist in a bewildering variety of designs. The advantage of most harnesses is that they spread the forces generated by pulling over a wide area of the body and, if to the chest, to a strong part of the body. Plainly, some designs encourage pulling, such as would be beneficial to tracking or sledge dogs. Other designs of harnesses place local compressive forces on the dog's axillae (i.e. beneath the forelegs). This is a painful effect that might cause bruising, even damage, to sensitive skin and to the underlying brachial nerve complex.

A more recent innovation in harness designs are front control models, where the dog is tethered or led from the chest (sternal region). Such designs have all the power steering advantages of a headcollar, but without the discomfort and restrictions previously mentioned.

Muzzles are another useful accessory for the management of dogs, simply preventing biting, self-mutilatory acral licks or ingestion of undesirable material (e.g. coprophagia). Muzzles that are to be worn for long periods of time should be of a size and design that permits panting, i.e. not restrict mouth gape. The fabric styles of muzzle with an open front restrict gape and so are only suitable to be worn for brief periods, such as during a veterinary examination.

Electronic training aids

There is considerable controversy concerning the use of devices which administer painful or aversive electric shock. There is wide country to country variation in their frequency of use for training dogs, indeed in their legality. Such devices are not permitted by law in Switzerland, Denmark and other countries, yet they can be freely purchased in the USA and some European countries, including the UK.

The advantage of say, radio-controlled electric shock collars is that the trainer can deliver a painful stimulus at the moment the dog is engaging in some inappropriate behaviour, such as chasing farm livestock. Difficulties arise when the shock intensity is poorly matched to the pain sensitivity of the dog concerned. Too painful and the dog will form superstitious fears about being in a field or near trees where it received the shock. Too low and the dog will continue to engage in highly motivated chase behaviour and will be desensitized to the painful stimulus.

More sophisticated paradigms might usefully be introduced to improve the effectiveness of such radio-controlled shock devices: pre-training the animal to respond to both a shock and a pre-shock tone in situations that are less exciting or less distracting. By an improved overall level of obedience, the trainer stands a better chance of interrupting such inherently rewarding behaviours as hunting.

The controversy surrounding use of shock collars continues to be divisive in the veterinary and behaviour therapy community, especially as it can now be well argued that there are effective alternatives. Dog training by the reward-based methods and including clicker training paradigms have made training more predictable in its outcome. Long distance interruption of unwanted behaviour can now be achieved using a gas release device, sometimes referred to as a citronella collar. These are designed to release a small interruptive spray of gas from a cylinder mounted under the dog's collar, perhaps at the moment it barks, steals food or chases cyclists.

Studies by Juarbe-Diaz and Houpt (1996) indicate that the effectiveness of the gas-training systems is greater than electric devices for interruption of unwanted barking, without the animal welfare objections of shock collar use.

Innovation in design and imagination characterize the use of equipment for behaviour modification. On the one hand, many dog owners want 'quick fixes' and have neither the skill nor the inclination to apply complex behaviour modification or retraining regimes. Such clients usually take well to gadgets and are grateful for technological fixes. However, significant numbers of owners of problem pets take a more spiritual approach which recognizes the need to understand the dog's fundamental needs and emotions, and are willing to acknowledge that there must be changes in the relationship.

Pheromones

In the last decade, there has been an interesting area of applied research led by French veterinarian Patrice Pageat (1999). He has investigated the composition of sebaceous secretions of the lactating bitch and subsequent responsiveness of adult dogs to a synthetic analogue of the same secretions. Pageat named this 'dog appeasing pheromone', on the basis that it seems to reduce anxiety or fear states in dogs. The product DAP has been marketed worldwide with persuasive clinical trials to establish efficacy for treating both chronic and acute anxiety states. It seems to be particularly helpful in the management of fear when the dog is confined to the house, where the DAP can be contained and reach a higher concentration than outdoors. A collar-version of DAP is now available in certain countries for outdoor applications.

The question arises whether this is a genuine pheromonal effect, implying that the dog's reactions are inborn or biologically determined rather than learned. There are numerous definitions of the term pheromone (see Doty, 2004), and there is controversy about its implied invariable or reflexive effects, even in the world of insects. This writer believes that the DAP effect (which is genuine though modest) is simply a reminder of how things were when the puppy was contentedly suckling his or her mother. In this respect, it is like the powerful reminiscing effect of perfumes and other odours which evoke distinctive memories in humans.

Environmental enrichment

Professor John Webster (1994) introduced the concept of five fundamental rights or needs of animals as freedom from hunger, thirst, discomfort, fear and freedom to mix with their own kind. It will be apparent to the reader that many dogs are exposed to fear, are prevented from mixing with other dogs and indeed are made to spend long periods in stressful social isolation. In the world of farm and zoo animals, the concept of environmental enrichment is popular, where the physical and temporal world of the animal is made more interesting, more varied and less predictable. For instance, primates may have to use a tool in order to extract termites from a hole in wood, or pigs eat their food over an extended period by having food dispensers or food offerings in frequent short meals rather than a single, large gorged meal.

In the world of dogs, devices that similarly slow down the rate of consumption of food, or force the dog to work for his food, have been introduced. One such is the Buster Cube (Kruuse DK); another the rubber Kong toy, which may be stuffed with food. The significance of these devices is greater where a contemporary lifestyle leads to dogs being left alone for very long periods (see earlier). In addition, modern, processed foods can be eaten very rapidly by the dog, denying it the opportunity to engage in predatory/hunting activities and slow mastication of a varied diet.

The most important feature of a dog's welfare is that it has the company of other dogs. Where the keeping of pairs or trios of dogs is unaffordable or not practical, a desirable alternative is that the pet be taken to a dog day care centre, which is a popular option in North America, though still unusual in Europe.

Concluding Remarks

Most behavioural problems have their origin in actions by the owner, by the owner's habits and by the environment which they provide for the dog. A few problems arise due to medical or genetic factors. There has been a marked improvement in the outlook for the treatment of behavioural problems in dogs due to an explosion of knowledge about canine social behaviour, also due to the increased sensitivity of veterinarians in practice to the importance of canine behaviour. The breakdown rate of relationships between man and dog varies between countries, but in the UK is of the order 10% per dog lifetime. This suggests that 500,000–700,000 dogs in the total population of approximately 7–8 million will have been rehomed, many of them for behavioural reasons. The man–dog relationship is, however, more secure and longer lasting than the institution of marriage, which has an only 50% 'survival' rate in many Western countries. Despite, or perhaps because of, their misbehaviours, dogs evoke remarkable loyalty from their people!

References

Billinghurst, I. (nd) drianbillinghurst.com
Doty, R. (2003) Handbook of Olfaction and Gustation. Marcel Dekker, New York, p. 345.
Harvey, M.J.A., Cauvin, M., Dale, M., Lindley, S. and Ballabio, R. (1997) *Journal of Small Animal Practice* 38, 336–339.

Juarbe-Diaz, S. and Houpt, K. (1996) Comparison of two antibarking collars for treatment of nuisance barking. *Journal of the American Animal Hospital Association* 32, 231–235.

Mugford, R.A. (1987) The influence of nutrition on canine behaviour. *Journal of Small Animal Practice* 28, 1046–1055.

Overall, K.L. (1997) *Clinical Behavioral Medicine for Small Animals.* Mosby, Missouri.

Pageat, P. (1999) Attachment and pheromones in the dog. *Proceedings of 2nd World Vet Meeting on Veterinary Ethology.*

Polsky, R.H. (1994) Electronic shock collars: are they worth the risks? *Journal of the American Animal Hospital Association* 30, 463–468.

Pryor, K. (1975) *Lads Before the Wind.* Sunshine Books, Waltham, Massachusetts (revised 2000).

Shepherd, K. (2002) Development of behaviour: social behaviour and communication in dogs. In: *BSAVA Manual of Canine and Feline Behavioural Medicine.* BSAVA, Gloucester.

Sheppard, G. and Mills, D.S. (2003) Evaluation of dog-appeasing pheromone as a potential treatment for dogs fearful of fireworks. *Veterinary Record* 152, 432–436.

Sigler, L. (1991) Pet behavioral problems present opportunities for practitioners. *American Animal Hospital Association Trends* 4, 44–45.

Stead, A.C. (1982) Euthanasia in the dog and cat. *Journal of Small Animal Practice* 23, 37–43.

Tuber, O.S., Hothersall, O. and Voith, V.L. (1974) Animal clinical psychology: a modest proposal. *American Psychologist* 29, 762–766.

Webster, J. (1994) *Animal Welfare: A Cool Eye Towards Eden.* Blackwell Science, Oxford, UK.

14 Behaviour and Disease in Dogs

Åke Hedhammar and Karin Hultin-Jäderlund

Introduction

Interactions between behaviour and disease have several dimensions. In many cases where dogs have behaviour alterations, the behaviour is integrated with signs of a physical disease.

By definition, specified behavioural disorders are, as other diseases, denoted by a name for the specific condition; a diagnosis. Such diagnoses of abnormal behaviour are commonly *symptomatic*, e.g. convulsion and pica 'just' describing the signs. Diagnoses at a higher hierarchical level should be more significant, functional and related to aetiology, e.g. convulsions due to a known infectious agent causing an inflammation in a certain part of the brain as in rabies, or pica due to malnutrition by a pancreatic insufficiency due to an inherited atrophy of the pancreas as described in German shepherds and collies. Ideally diagnoses at a higher hierarchical level, i.e. *functional* diagnoses, are defined at a molecular level as, for example, narcolepsy due to known and sequenced genotypes.

Contrary to narcolepsy caused by single gene mutations, most behavioural alterations supposedly are multifactorial and involve many genes as well as environmental factors. Genes as well as environmental factors, including socialization and training, altering behaviour are the focus of other chapters in this book. In this chapter we will try to elucidate the interactions between behaviour and physical diseases affecting behaviour. The disease on the other hand might be genetically determined as well as affected by environmental exposure. Even post-traumatic stress due to disease, or interventions to diagnose or treat a disease, are examples of interactions between behaviour and disease that have practical implications in a clinical setting.

Besides specified behavioural diseases in which an abnormal behaviour is the only complaint, almost all other diseases manifest partly in behavioural alterations. In fact, owners of a sick dog most commonly react on behavioural rather than physical changes. When recalling the clinical history of a case – the *anamnesis* – owners almost exclusively bring up changes in activity level and changes in eating, drinking and eliminating (urinating, defecating) behaviour.

In many diseases behavioural alterations are significant findings aiding diagnosis at a higher hierarchial level. For example, polyuria (pu) – excessive urination – is a common and unspecific symptomatic diagnosis, but polyuria in combination with lethargy, polydipsia (pd) and polyphagia is a strong indication for a more specified diagnosis – Cushing's disease – a disease with an increased release of glucocorticoids from the adrenal gland.

In skin diseases, the clinical sign of itch has a major importance differentiating many diagnoses from each other.

Behavioural signs are no exception to other clinical signs in that there is no strict border between normal and abnormal. Behaviour may also be inappropriate rather than abnormal in many cases.

Many diseases result in profound behavioural alterations. Pain accompanying the course of several diseases may elicit aggression and anxiety to a great extent. It must also be appreciated that pain and distress can also be related to diagnostic procedures and treatments for many diseases.

In this chapter, we will define a terminology used in small animal veterinary medicine that helps to understand an interaction between canine behaviour and canine diseases. We will then bring attention to clinical signs/symptoms related to behaviour and disease, and give examples from diseases in various organ systems and of different aetiologies. Epilepsy, pain and signs of anxiety will be given special attention.

The attention to canine behaviour in a clinical setting will be put into focus and clinical signs related to specific syndromes affecting canine behaviour will be reviewed.

Definitions

Anamnesis – recalling the history of events preceding and during the course of an illness.
Anxiety – 'experiencing' a sense of dread or fear (especially of the future) of real or imagined threat to one's mental or physical well-being.
Cognitive dysfunction – change in interactive, elimination or navigational behaviours, attendant with ageing, that are explicitly not due to primary failure of any organ system.
Convulsion – primarily a lay term. Episodes of excessive, abnormal muscle contractions, usually bilateral, which may be sustained or interrupted.
Coprophagy – the eating of excrement.

Disease – literally 'lack of ease', a pathologic condition of the body that presents a group of clinical signs peculiar to it (and which set the condition apart as an abnormal entity differing from other normal or pathological body states).
Epilepsy – a chronic neurological condition characterized by recurrent epileptic seizures.
Epileptic seizure – manifestation(s) of epileptic (excessive and/or hypersynchronous), usually self-limited activity of neurons in the brain.
Fatigue – a 'feeling' of tiredness or weariness resulting from continued activity.
Fear – the emotional reaction to an environmental threat.
Itch (pruritus) – an unpleasant sensation that causes an individual to rub, lick, chew or scratch at its skin.
Listlessness – too tired to show an interest.
Narcolepsy – a disorder of the brain that is marked by sudden recurring attacks of sleep.
Pain – an aversive sensory and emotional experience (a perception) which elicits protective motor actions, results in learned avoidance, and may modify species-specific traits of behaviour, including social behaviour.
Pica – a perversion of appetite with craving for substances not fit for food.
Restlessness – unable to rest.
Seizure – non-specific, paroxysmal, abnormal event of the body (cf. epileptic seizure).
Weakness – lack of physical strength.

Clinical Signs Related to Behaviour and Disease in Dogs

Level of consciousness

Altered states of consciousness are always related to abnormal brain function. The following states of consciousness should be taken into account:

- *Normal* – the dog is alert, responds to external stimuli, is aware of its surroundings, and responds to commands as expected.
- *Depressed* – the dog responds slowly or inappropriately to verbal stimuli. Some animals may appear disoriented or even act delirious. Most sick animals are depressed.
- *Stuporous or semi-comatose* – the dog is generally unresponsive and appears to be asleep, but can be aroused by strong stimulation, especially pain.
- *Comatose* – the dog is unconscious and no behavioural reactions are possible to elicit by any stimulus.

A dog that is semi-comatose or comatose most commonly has either a disease process or a lesion of traumatic origin affecting the brain stem. A dog in that state might also be affected by some toxin or by any metabolic disease secondarily affecting the brain. Less commonly, a semi-comatose or comatose dog has a diffuse, bilateral cerebral disease.

Narcolepsy

Narcolepsy (sleep attacks) with concomitant cataplexy (loss of muscle tone in all skeletal muscles except muscles needed for breathing) is an uncommon clinical sign occurring in dogs. The signs in themselves come from an imbalance in the sleep–awake neurotransmitter system in the forebrain and brain stem. In the dog population, narcolepsy with cataplexy occurs both as a genetic disease and as a sporadic acquired disease. Even though the clinical signs in different dogs suffering from different forms of the disease are undistinguishable from each other, the cause of imbalance in the affected neurotransmitter system seems to differ. In the hereditary genetic form, the receptors on the postsynaptic nerve cells are mutated and malfunctioning, while in the acquired form, the production of the neurotransmitter substance seems to be too low.

Weakness – tiredness – fatigue – listlessness

These are signs most commonly related to physical diseases in dogs. Weakness is a sign commonly encountered in small animal veterinary practice. It is termed episodic when elicited by exercise and dissipated by rest. Although weakness unrelated to exercise is an appreciated clinical sign in human psychiatry, it is not easy to reveal such aetiology in canine weakness. As several metabolic, endocrine, cardiovascular and neuromuscular diseases may result in weakness, work up of a weak dog calls for an evaluation of electrolyte and hormonal balances as well as cardiovascular and neuromuscular function. Adrenal insufficiency (Addison's disease), hypothyroidism and Myasthenia gravis are examples of differential diagnoses in a dog that develops unexpected severe weakness at almost any age. Weakness at an older age could also be due to degenerative processes, including canine cognitive dysfunction.

Anxiety – fear – avoidance – flight

These signs might be caused by injuries and physical diseases. Although anxiety and fear are commonly seen as inherent behavioural manifestations in many dogs, it must be appreciated that injuries and physical diseases may cause or perpetuate signs of anxiety and fear. Avoidance and flight in a dog that has not exhibited such signs before indicate a possibility that an ongoing disease process or current or earlier exposure to physical harm might be involved. Any traumatic injury, causing pain somewhere in the body, may provoke anxiety and fear. But also to 'experience' an epileptic seizure or loss of breath is capable of inducing anxiety and fear. Anxiety is also described as an early clinical sign in lysosomal storage diseases such as fucosidosis, an inherited disease in English springer spaniels, for example. Relief of pain before and after operative procedures has received increased attention in veterinary medicine. In order to recover fast and completely, pain relief is almost as important as anaesthesia during surgery.

Aggression

Aggressive behaviour might also arise out of external 'stimuli'. Misdirected aggression cannot be excused but rather explained by painful processes that the dog may either relate to a person or to a situation. Clinical situations and veterinary visits might either be so harmful that aggression is warranted or just be a situation in which a dog with strong dominance aggression 'tests' his position. Sudden and strong pain may elicit biting that is an expression of avoidance rather than aggression. Relief of pain is as important for reduction of misdirected aggression as for anxiety.

Epileptic seizures and epilepsy

Epilepsy (recurring epileptic seizures) is one of the most common symptomatic diagnoses arrived at in dogs in veterinary practice, even though the veterinarian seldom observes any clinical signs in a dog with epilepsy. Attacks described by the owner lasting a couple of minutes most often are easily recognized as epileptic seizures, especially when general muscle activity (cramping) appears in the seizure. Another feature most often recognized in conjunction with an epileptic seizure is the so-called postictal period, a transient clinical abnormality of central nervous system function that appears when clinical signs of the epileptic seizure have ended. The postictal phase may last for a few minutes to several hours, during which the dog may be restless, lethargic, confused, disoriented, aggressive and/or blind. A problem in reaching a symptomatic diagnosis in some dogs might be when the muscle activity is not generalized, when the dog is conscious during the seizure, when there is no postictal phase and/or when the seizure is mainly sensory.

Definitions about epileptic phenomena are all in accordance with the glossary from the International League Against Epilepsy, an international commission synchronizing the platform on which knowledge about epilepsy in humans is gathered. Since the inherent neurophysiological and neuroanatomical bases for epileptic seizures are similar in all mammals, comparative aspects are useful when dealing with epileptic dogs. However, in human medicine, electroencephalography (EEG), the graphic recording of the spontaneous electrical activity of the brain, has a vital importance in patients that have experienced seizures. EEG-waves from neurons discharging during an epileptic seizure have typical, abnormal patterns. In the clinical setting, this might be achieved by constant EEG-registration with simultaneous video-registration of the patient's behaviour until a seizure has been experienced. However, in dogs this technique has major concerns in a clinical setting, e.g. due to artefacts from general muscle activity.

EEG-registration from dogs experiencing seizures have been performed at some university clinics, with the animal anaesthetized or sedated to prevent muscle artefacts – but at the same time affecting the neuronal activity – and for shorter periods (not expecting a seizure during the registration). Sensitivity and specificity of EEG-abnormalities during those circumstances have, so far, not been very high.

In cases where dogs have had seizures, not easily recognized as *epileptic* seizures from the owner's history or from looking at videos of the event, a simple test confirming the event as epileptic is still lacking. Examples of events that could be mixed up with epileptic seizures of different kinds are syncopes, sleep attacks in narcolepsy, and compulsive disorders (see also Chapter 13 in this volume).

An epileptic seizure is always a sign, either of a low or a decreased seizure threshold, or of a lesion present in the cerebral cortex, containing hyperactive nerve cells. The term 'seizure threshold' means the resistance in the nervous system against developing seizure activity. When seizure threshold is critically low, an epileptic seizure occurs. The level of the seizure threshold varies interindividually as well as with age. Precipitating factors temporarily lowering the threshold could include elevated body temperature.

Epileptic seizures occurring in a dog should initially be looked upon as unspecific signs of an underlying disease. Epileptic seizures are elicited from neurons in the grey substance of the brain. All kinds of diseases in the forebrain (e.g. inflammations, tumours, anomalies, traumatic injuries) might give rise to epileptic seizures. Also diseases of internal organs, or intoxications, altering the metabolic state in the body and thereby the brain function, may present as seizures. Common examples of that include liver diseases, hypoglycaemic conditions and electrolyte disturbances. Conventional blood and urine analyses are often sufficient to confirm these kinds of underlying diseases.

In some cases, no matter how extensive examinations are performed with various diagnostic aids, no metabolic changes are reflected in blood or urine and no morphologic or pathologic changes can be detected in the brain. The affected dog is then said to have a 'low seizure threshold'. The ultimate diagnosis often used for these cases is 'idiopathic epilepsy'. The threshold for seizures may be an inherited trait. In some breeds, inherited forms of epilepsy are known to occur (e.g. golden retriever, labrador retriever, Bernese mountain dog, keeshond, Belgian tervueren, Shetland sheepdog and vizsla). However, there are also sporadic cases of idiopathic epilepsy that could appear in dogs of any breed in any breeding line.

In an effort to reach an aetiologic diagnosis in a dog with epileptic seizures, the practising veterinarian has to act as a detective. The work-up should include a thorough anamnesis, a clinical and a neurological examination of the patient. Blood and urine samples should be analysed regarding possible metabolic/endocrinologic/toxic aetiologies. Further investigations might be indicated for certain cases, where magnetic resonance imaging (MRI) of the brain and liquor tap analysis are especially valuable diagnostic aids. In the future, molecular genetic analyses regarding different kinds of inherited epilepsy among affected breeds hopefully will come in place.

Pruritus (itch)

Itch is a clinical sign that is most commonly related to skin diseases by external stimuli such as ectoparasites or immunological conditions such as atopic dermatitis. At the far end of a scale describing intensity of itch, self-mutilation may occur. Self-mutilation might be associated with a sensory neuropathy seen as an inherited condition in pointers and longhaired dachshunds at a young age, where sensory nerve cells are malfunctioning, probably causing paresthesias (Gnirs and Prelaud, 2005). Acquired sensory neuropathies with self-mutilation can also be seen in any breed due to inflammatory disease of the cranial and spinal ganglia and dorsal nerve roots (canine ganglioradiculitis) or by ingesting infected meat from pigs with Aujeskies disease (pseudorabies). The latter is a subacute viral brain infection and the former has a more chronic manifestation with Siberian huskies as an example of an over-represented breed. Also peripheral nerve tumours and the so-called paraneoplastic syndrome may result in peripheral sensory neuropathy and self-mutilation. Intensive licking, for whatever reason, may result in a skin disease named acral lick dermatosis.

Pain

Pain undoubtedly affects behaviour. Since pain is included in the clinical signs of a wide variety of different diseases and injuries in dogs, it is an important clinical sign to recognize – and to relieve. The main categories of painful processes in dogs emanate from localized inflammatory reactions, traumatic injuries, disc-associated diseases or some neoplastic diseases.

Some different patterns of behaviour might be recognized for a dog in pain:

- A stiffening of its body posture, a whine or a growl, a turning of the head towards the stimulus, or a bite when we touch a certain part of its body, could be the behaviour telling us that a dog perceives pain from being touched there. These kinds of reactions are seen in cases such as localized subcutaneous abscesses. One could also elicit these reactions by causing pain in an unaffected part of a body, e.g. by giving an injection.
- A dog also avoids painful movements, if possible, of the body part involved in pain. This could be noticed as a lameness of a fractured limb, or chewing of all food on one side of the mouth when a tooth root on the other side is inflamed.
- A dog that is reluctant to move, stiff while having to move, avoiding certain body positions, screaming spontaneously while changing posture and maybe also screaming out loud as soon as *any* part of its body is touched, has behaviours that should be interpreted as emanating from severe pain. The painful disease or injury is then often found somewhere in the trunk and especially in the spine. In the authors' experience, the diagnoses most often found in dogs showing really severe pain are disc prolapses, meningitis, skeletal malignancies or panosteitis (a painful disease process in the skeleton during growth).

Even though a dog in pain might bite a person touching (provoking) it, unprovoked aggressiveness in dogs is most often not related to painful diseases or injuries.

Dogs that experience a sudden paralysis of their hind limbs due to some non-painful event, e.g. an infarct in the spinal cord, can be very restless and anxious the first days after that event, constantly moving around and changing postures. This behaviour is not typical for a dog in pain, but is sometimes confused with that.

Appetite

Normal and even 'good' appetite is a clinical sign accompanying good health. Decreased appetite in a dog normally eating well is indicative of disease and/or unease. Organic diseases of various aetiologies have an impact on appetite. Whether decreased appetite can be related to mental depression in dogs is quite difficult to evaluate. Many dog owners report decreased appetite in circumstances when other dogs or members of a household are absent.

Conditions similar to anorexia nervosa have not been proven to occur in dogs, although lack of appetite for no obvious reason has been noted in sledge dogs, for example. Ravenous appetite can be elicited by endogenous corticosteroids as in Cushing's syndrome and by exogenous supply of corticosteroids in treatment protocols for immunological diseases. In those cases it is also accompanied by polydipsia and polyuria.

Inappropriate urination

Urination in inappropriate environments can be due to diseases resulting in an increased urinary volume (polyuria), congenital defects, inflammations or difficulties controlling bladder function (incontinence). 'Voluntary' voiding of urine at inappropriate places in housebroken dogs is a behaviour that has to be differentiated from involuntary voiding of urine in a dog with urinary tract disease, excessive urinary production or neurological disease.

The first step is to find out whether it is related to an increased urinary volume – polyuria (pu). Polyuria is most commonly seen together with an increased water intake – polydipsia (pd) – that makes it difficult for the dog to refrain from urinating. Pd/pu can be due to congenital diseases in the liver or kidneys in young dogs, which has to be watched for in puppies that will be unable to be housebroken. In older dogs that have not exhibited inappropriate urination behaviour before, there are several causes of pd/pu, e.g. chronic liver failure and renal diseases, parathyroid tumours, lymphosarcoma, diabetes, Cushing's syndrome (overproduction of corticosteroids) or pyometra.

Compulsive water drinking – pseudo-psychogenic polydipsia – also resulting in pd/pu might arise from emotional stress and will respond to water deprivation contrary to most diseases. A dog with Cushing's syndrome also responds to water deprivation but in contrast to the dog with pseudopsychogenic polydipsia, in addition exhibits polyphagia (see below, Endocrine system).

Having ruled out pd/pu as the cause of inappropriate urination, the next step is to rule out congenital abnormalities triggering the micturation process, such as an ectopic ureter or pelvic bladder, or urinary tract irritation from inflammation or bladder calculi.

Difficulties in controlling bladder function resulting in incontinence might arise from diseases and injuries of the nervous system as well as within the lower urinary tract. In male dogs, diseases of the prostate also might affect the micturation.

To differentiate between physical diseases and inappropriate urination for other reasons, water intake and urinary volume should be measured. A daily water intake between 20 and 90 ml/kg bodyweight is considered normal and a daily urine output between 20 and 40 ml/kg bodyweight is to be expected. Intervals and frequency of 'normal' urinary voids varies greatly between and within individuals. Changes over time are often more revealing. Voiding when left alone or at night is difficult to watch. A dog with urinary incontinence for medical reasons usually voids urine in the bed rather than spread over the home. Urine analyses including specific gravity, protein, glucose, cellular components and bacteriological cultures help to differentiate between different medical causes as do blood analyses including glucose, electrolytes and white blood cell counts.

Inappropriate defecation

Defecation at inappropriate locations is most commonly associated with organic diseases. Diarrhoea, defined as an increased volume and water content resulting in an increased frequency of defecations, may also result in defecation at inappropriate places. An expected amount of normally formed stool at an inappropriate place is more likely to be due to an elimination behaviour problem than stools consistent with enteritis. Alternatively, loose stools may indicate physical or psychological stress as that speeds up the passage of food through the digestive tract with resultant diminished time for water reabsorption in the colon.

Diseases Affecting Behaviour in Dogs – By Organ System

Almost all diseases can affect behaviour to a variable degree, in dogs as well as in humans. One major difference is that dogs cannot 'worry' about the outcome of serious and life-threatening diseases. Nor do they have 'insight' of the potential of later relief to help cope with pain and distress. As an overview, common behaviour alterations in particular diseases are presented below by organ system. The overview begins with effects from diseases in the nervous system and sensory organs.

Nervous system

Clinical signs from diseases in the nervous system include altered behaviour in one way or another. The possible underlying categories of diseases are often, in clinical neurology, systematized as follows:

- Vascular diseases (infarcts or haemorrhages affecting the nervous system).
- Inflammatory/infectious/immunological diseases (e.g. distemper, steroid responsive meningitis-arteritis).
- Traumatic injuries (e.g. skull fracture).
- Anomalies (e.g. hydrocephalus).
- Metabolic/toxic/nutritional disturbances (e.g. hypoglycaemia, strychnine poisoning, thiamine deficiency).
- Idiopathic conditions.
- Neoplastic diseases (e.g. primary brain tumours, metastatic tumours in the central nervous system, CNS).
- Degenerative diseases, disc-associated (e.g. disc prolapse).
- Degenerative diseases, degeneration of nerve cells (e.g. hereditary ataxia in fox terriers).

In many of these diseases, in addition to the pure 'behavioural' component (e.g. becoming more timid, or aggressive, or extremely restless, or not recognizing the owner any more), signs such as blindness, paralysis, paresis, exaggerated movements, uncoordinated movements, tremor, in whole or part of the body present concomitantly. Pure 'behavioural' components as well as seizures are strongly correlated with pathological conditions in the forebrain.

Also diseases in the peripheral nervous system alter behaviour, even though the brain cells are unaffected by pathology. For example, a dog with paretic jaw muscles due to an inflammation in the motor nerves to the masticatory muscles shows great difficulty in eating and drinking since it is unable to close its mouth. Another example is a dog that becomes urinary incontinent due to a tumour growth affecting nerves in the pelvic region, innervating the urinary bladder and urethra.

Meningitis
One quite common form of meningitis in dogs, steroid-responsive meningitis-arteritis, occurs in young adult dogs, and immune-mediated mechanisms are strongly suspected to cause the disease. The disease seems to be more common in some breeds, e.g. boxers and Bernese mountain dogs. Main clinical signs are neck stiffness, spinal hyperestesia, and fever. Clinical signs are promptly relieved by corticosteroid therapy in most cases.

Brain tumours
The clinical signs of brain tumours depend on the location of the mass inside the brain. One common sign from tumours growing anywhere in the forebrain is

seizures. Tumours in the frontal or temporal lobe often also cause other abnormalities in behaviour, such as aggressiveness, extreme restlessness, confusion or an inability to understand instructions anymore. Another behavioural sign connected with diseases of the forebrain, often a brain tumour, is so-called compulsive walking – an aimless continuous wandering. Brain tumours in the occipital lobe often cause visual field deficits. Tumours in the pituitary gland and/or hypothalamus in addition also may cause autonomic and endocrine signs: polyuria, polydipsia, changes in eating and sleeping pattern, and so forth. Main signs of tumours in the brain stem are gait deficits and cranial nerve signs. In later stages of brain tumours, significant alterations in consciousness should be expected.

Sensory organs

Deafness, blindness, inability to smell or taste, inability to perceive pain and/or touch and dizziness due to an inability to perceive influences on the balance organs of the inner ear are all clinical signs, occurring with certain diseases seen in dogs and profoundly altering the behaviour of the affected dog.

Deafness
Bilateral deafness for a dog means a dependence on other sensory organs for interactions with the surroundings. As long as the visual system in particular is intact, and the head is turned towards an interacting event, an observer might miss that the dog has any deficit at all. The deficit becomes evident as soon as the dog turns its head away from anyone trying to communicate with it. Unilateral deafness is not always revealed clinically, but is noticeable when the dog can hear but is unable to localize the sound. Deafness in young dogs is most often a congenital disease, proven or suspected to be inherited in different affected breeds. Border collie, cocker spaniel, collie, dalmatian and doberman pinscher are some breeds with reported congenital deafness. When elderly dogs become deaf, a loss of cells in the auditory system should be suspected.

Blindness
Blindness comes from either diseases in the eyes (e.g. glaucoma, retinal detachment), or diseases affecting the central nervous system visual pathways (e.g. hydrocephalus, hepatic encephalopathy, lead intoxication, inflammatory disorders of the optic nerve). A dog that has slowly and progressively become blind, and is kept in surroundings that have not changed since the disease started, might behave in a surprisingly normal manner. But as soon as the owner moves furniture around in the home, or moves to another place with the dog, the deficit becomes evident. A dog with acute onset of blindness will show obvious behaviour alterations, often manifested as an abnormally cautious behaviour and walking into objects.

Inability to smell
Inability to smell is occasionally a complaint from owners of hunting dogs. This clinical sign could be due to infestation by *Pneymonyssus caninum* in the nasal cavities, for example.

Inability to perceive pain and/or touch
Inability to perceive pain and/or touch can be related to severe injuries of the spinal cord, dorsal nerve roots or peripheral nerves. These injuries also, in almost all cases, concomitantly lead to paralysis of the affected body part. In addition there are some breed-related, possibly inherited, diseases with loss of function in the sensory nerves as the pathologic condition. Complete loss of pain/skin sensation in dogs can give rise to severe self-mutilation, e.g. eating its own tail or toes.

Dizziness
Dizziness, from disturbances in the vestibular system (balance system) is seen quite commonly in dogs. Signs that might be included in the resulting 'vestibular syndrome' are head tilt, uncoordinated movements, jerky eye movements and strabismus. Diseases that could be involved include deep ear inflammation, intoxication, thyroid disease and neoplasia.

Endocrine system

Next to altered behaviour from diseases affecting the nervous system, altered behaviour due to diseases in the endocrine system might be the most common and obvious to the owner. By including effects from altered levels of hormones regulating reproduction it is a clinically highly relevant group of diseases affecting mainly eating, drinking, elimination and sexual behaviour.

Variations in sexual hormone production and release due to castration (orchiectomy in male dogs and spaying in bitches) are outside the scope of a relation to diseases unless performed for medical reasons. *Pyometra* (a manifest infection in the uterus) most commonly is cured by ovarihysterectomy, so also affecting the production and release of oestrogens and progesterones. Chronic diseases affecting the prostate but also perineal hernias and perineal skin diseases are favourably affected by orchiectomy. Dogs castrated for medical reasons are most commonly older than those castrated for non-medical reasons. An anticipated positive effect on behaviour in dogs castrated for non-medical reasons is less pronounced and not even desired when castrated for medical reasons. Spayed bitches are more prone to urine incontinence, which has to be differentiated from spraying. Old dogs that have had an orchiectomy will roam to a somewhat lesser degree and dominance aggression may diminish, but effects are unpredictable and variable.

Occasionally bitches with ovarian tumours exhibit behavioural alterations, but rarely nymphomania. Male dogs with *sertolicell tumours*, usually in a retained testicle, exhibit 'feminization' of their behaviour together with gynaecomastia (enlarged mammary glands) and flaccidity of the preputium. Older chryptorchid dogs in particular should be watched for this when exhibiting altered behaviour. Painful processes in the prostate might give rise to withdrawal or aggression as well as limping.

Hypothyroidism, the most common endocrinopathy in dogs, has been indicated to cause various alterations in behaviour, including aggressiveness. There are figures from the 1980s indicating that less than 2% of aggressive behaviour was due to hypothyroidism, indicating that it is not as common as was earlier thought. Lethargy and weakness are signs related to a decrease in the metabolic rate in hypothyroidism. Whether it also might give rise to mental depression is difficult to evaluate in dogs. Hyperthyroidism, which has been claimed to cause aggressiveness in dogs, is an extremely rare condition and it has not been possible to prove this a clinical cause of aggressiveness.

Hyperadrenocorticism (Cushing's syndrome) has a profound effect on metabolism and includes such mental signs as listlessness. The most prominent clinical effect on behaviour is its effect on eating and drinking. It results in polydipsia (pd)/polyuria (pu) as well as polyphagia. It is usually a disease of older age and has a gradual onset of clinical signs.

Hypoadrenocorticism (Addison's Disease), on the other hand, may also occur at quite a young age and have either a gradual and diffuse clinical onset or may result in sudden episodic weakness due to hyperkalaemia. There are fairly strong breed predispositions for Addison's disease in bearded collies and standard poodles, warranting attention to diffuse mental changes in these breeds. However, it may be seen also in other breeds.

Diabetes mellitus is not only a cause of pd/pu and polyphagia but also muscle weakness due to disturbed electrolyte balance and other complications.

Gastrointestinal system

Severely affected exocrine pancreas function, as can be seen in German shepherds and rough collies with an inherited disease causing pancreas atrophy, may have an impact causing weakness. Occasionally it may also result in coprophagy.

Malabsorption from severe diarrhoea and protein loss 'leaking' from the gut may result in similar effects on general behaviour and inappropriate eating habits. Inflammatory bowel diseases and severe colitis of various aetiologies may 'irritate' to such an extent that behaviour alterations such as anxiety and aggression may develop.

Liver diseases of various aetiologies may provoke pd/pu as well as weakness.

Hepatic encephalopathy (HE), most commonly caused by congenital portal vein anomalies or toxicity in young dogs and hepatic cirrhosis and neoplasia in older dogs, may provoke subtle behavioural complaints including pacing, aggression, hysteria, intermittent deafness and depression. The clinical signs in HE are caused predominantly by increased central brain ammonia concentrations due to urea cycle deficits.

Urogenital and reproductive system

Inflammatory processes of any part of the lower urinary tract may induce inappropriate elimination behaviour. Urinary calculi and enlargement of the prostate by hyperplasia or chronic infections may affect voiding of urine.

Cardiovascular system

Heart failure most commonly seen in older dogs, and mainly caused by endocardosis in small breeds and cardiomyopathy in larger dogs, can affect behaviour by causing unprovoked tiredness. In young dogs similar effects could be due to congenital heart defects such as aortic stenosis.

Canine behaviour in a clinical situation

Stress at a veterinary clinic might not only affect behaviour during the visit but could result in behavioural alterations in similar situations later on. It is not uncommon that shyness and even aggressiveness towards people is blamed on unprofessional handling at a veterinary clinic.

It could well be that stress by restraint or painful events at a veterinary visit is the first time a puppy has experienced such an event and is not necessarily the cause of similar behaviour later on. But, being a stressful event to many dogs (and their owners), veterinary visits should be organized in as positive a manner as possible.

Proper examination of an individual sometimes calls for restraint, and proper supervision could result in isolation in a ward very much different from what a dog is used to at home. Both circumstances could be frightening, whether associated with pain or not.

Surgery, even under general anaesthesia, can significantly affect any individual as do most medications whether by intention or through side-effects. Keeping that in mind helps staff at veterinary clinics to avoid, as much as possible, post-traumatic stress disorders in their patients.

Panic is evidenced by very strong immediate reactions such as increased heart rate, salivation, trembling, flushing skin and a shortness of breath. However, fainting is rare in dogs suffering panic.

By optimal arrangements all the way through the reception area, waiting room, examination rooms, surgical theatres and wards, some stress could be relieved for patients as well as their owners. Space and noise are parameters to work with. Of greatest importance is handling by the personnel involved. A firm but gentle handling and a positive voice in all situations help anxious as well as aggressive dogs to cope with procedures that have to be performed (Fig. 14.1).

It might also be pertinent to question whether a procedure has to be performed or not – e.g. to measure rectal temperature in a dog acutely injured in the rectum might not be necessary or even relevant.

Whenever possible, and without interfering too much with the clinical findings and outcome, pain relief should be administered. In veterinary medicine as well as in human medicine, it is now well appreciated how beneficial that is. Pre and post medications to surgery are today good practice in our small animal hospitals.

Consultancies at veterinary clinics with behaviour as chief complaint

Behaviour is no exception to the value of a complete and thorough clinical history. By age, breed and sex of a patient, many clues can be reached and it does often help to limit the clinical work-up procedure. A dizzy young chihuahua puppy is much more likely to have a hydrocephalus as the cause than an aged St Bernard. A middle-aged standard poodle is on the other hand more likely to show episodic weakness by Addison's disease than a very young puppy of any breed. Complete demographic information also contains information on family members and other pets in the household.

The clinical history – an anamnesis – should include not only circumstances around the chief complaint such as: For how long? Gradual or sudden onset? When in relation to other circumstances as rest and exercises? It should also contain general questions as eating, drinking and elimination behaviour regardless of the chief complaint.

At a clinical examination for aggressiveness and shyness, any painful events should be watched for. Provoked pain by manipulation of the neck, spine and/or limbs or distress by abdominal palpation and pain at rectalization of the prostate in male dogs should be looked for.

Foreign bodies should be looked for in paws, ears and mouth. An elevated body temperature is of diagnostic value, e.g. in meningitis. Special procedures depending on clinical signs include clinical pathology of blood and urine, diagnostic imaging and endoscopy.

Clinical Signs Related to 'Specific' Behavioural Syndromes

'Compulsive disorders'

There are a number of recurrent episodic behavioural alterations in dogs, hard to classify according to current knowledge and available diagnostic techniques. Historically, neurologists have looked upon them as possible epileptic seizures evolving from the temporal lobe in the forebrain, while behaviourists have been prone to call them compulsive disorders. Just looking at the clinical manifestations of such disorders is not enough to judge who's right and who's not. The neurophysiology of the brain is complex enough to be able to function in such a way that the clinical signs seen could result from either. Ideally, an EEG-registration from the surface of the brain in a calm dog (not affected by drugs) just before, during and after an episode, should throw some light on the issue. Unfortunately, this is difficult to perform in a clinical setting.

Some of these 'syndromes' are recognized as symptomatic diagnoses, most often breed-related (see Table 14.1), three of which are described in more detail below.

Table 14.1. Some examples of behaviour signs associated with compulsive disorders, and some typical breeds showing these signs.

Behaviour symptom	Typical breeds showing symptoms
Fly catching	Cavalier King Charles spaniel
Spinning	Bull terrier, Staffordshire bull terrier
Tail chasing	Bull terrier, German shepherd, Australian cattle dog
Rage syndrome	English springer spaniel, English cocker spaniel
Chasing light reflexes	
Staring at shadows	Border collie
Flank sucking	Doberman pinscher
Star gazing	
Freezing	Bull terrier

Fly catching

The dog acts as if repeatedly watching, and then catching, imaginary flies. These episodes can last for minutes to hours, even to full-time occupation for different dogs. The phenomenon seems to be breed-related in Cavalier King Charles spaniels. Fly catchers have been ophthalmologically examined, excluding eye diseases. Anti-epileptic drugs have been used without success (DeLahunta, 1983), which contradicts the idea of this phenomenon in Cavalier King Charles spaniels being an epileptic event. (Although there are also individuals with seizures, known to be epileptic, that do not respond to anti-epileptic medication.) The current recommendation is to treat these dogs with behaviour modification and drugs that interact with the neurochemical balance (Rusbridge, 2005).

On rare occasions, a moment of fly catching is the initial phase of a seizure in a dog experiencing a secondarily generalized epileptic seizure, with convulsions. The event should then be regarded and treated as epileptic.

Tail chasing

Some dogs chase their own tails, or spin around in small circles, much too often to be judged as normal. In some individuals, the intensity of this behaviour leads

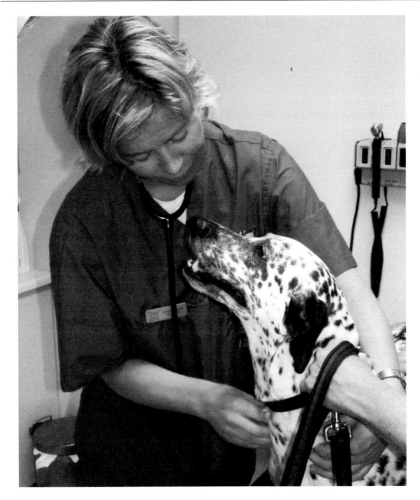

Fig. 14.1. Care and handling at a veterinary clinic is of outmost importance for canine behaviour as it relates to disease.

to dehydration and body weight loss because the affected dog doesn't take time to eat or drink. Dogs with this disorder should be clinically and neurologically examined to see if there are any clues for an evident aetiology for tail chasing or circling – e.g. inflammations of the anal sacs, or a morphologic brain lesion.

Many bull terriers seem to be predisposed to this behavioural disorder, and some scientific reports about the matter have been published. The disorder has been suggested by different authors to be an opioid-mediated stereotypy, a temporal lobe epilepsy or a compulsive behaviour. The latter two suggestions may not be mutually exclusive, as human beings with obsessive-compulsive disorders sometimes also have an associated seizure disorder. In one study, affected dogs had abnormal EEG patterns, indicating epileptic seizures, while unaffected control dogs had normal EEGs (Dodman et al., 1996). There are several treatment studies performed in dogs with this disorder, reporting responses to different kinds of therapy, i.e. anti-epileptics, anxiolytics, opioid antagonists, antidepressants and behaviour modification. The latest reports deal with antidepressants, e.g. clomipramine, and behaviour modification. Regardless of type of underlying brain dysfunction, it has been stated that there is a strong genetic component to these behaviours in the bull terrier breed.

Rage syndrome
Rage syndrome, also called idiopathic aggression or episodic dyscontrol, describes a behavioural disorder where an otherwise healthy, nice and friendly dog has recurring attacks of aggressiveness – often directed against members of the owner's family. There are cases reported with EEG evidence of temporal lobe epilepsy and response to antiepileptic medication, as well as other reports of cases with inconsistent EEG findings and disappointing therapy with anti-epileptics pointing to this disorder being 'just' a behavioural problem.

A veterinarian facing a patient with a history of sudden, unexpected, aggressive attacks, should perform a clinical and neurological examination together with blood and urine analyses, to check for possible underlying morphologic, metabolic or endocrine diseases. If not so, one should consider the option of this being rage syndrome, whatever that means.

Special attention to this syndrome has been paid in certain breeds, e.g. springer and cocker spaniels. Podberscek and Serpell (1996) found some evidence that this condition in the English cocker spaniel breed is an expression of social dominance rather than being a separate or pathological phenomenon. The clinical picture of rage syndrome should not be mixed up with the 'normal' aggressiveness included in the postictal phase of some epileptic dogs (with more classical epileptic seizures, cf. earlier text in this chapter). Those dogs are in a stage after their epileptic seizure where they have no chance to obey or control their behaviour. One should bear in mind that dogs with unpredictable or uncontrolled aggressiveness are potentially dangerous to people and other animals in their surroundings.

References and Suggested Further Reading

DeLahunta, A. (1983) *Veterinary Neuroanatomy and Clinical Neurology*. W.B. Saunders, Philadelphia, Pennsylvania.

Dodman, N.H., Miczek, K.A., Knowles, K., Thalhammer, J.G. and Shuster, L. (1992) Phenobarbital-responsive episodic dyscontrol (rage) in dogs. *Journal of American Veterinary Medical Association* 201, 1580–1583.

Dodman, N.H., Knowles, K.E., Shuster, L., Moon-Fanelli, A.A., Tidwell, A.S. and Keen, C.L. (1996) Behavioral changes associated with suspected complex partial seizures in Bull Terriers. *Journal of American Veterinary Medical Association* 208, 688–691.

Gnirs, K. and Prelaud, P. (2005) Cutaneous manifestations of neurological diseases:

review of neuro-pathophysiology and disease causing pruritus. *Veterinary Dermatology* 16, 137–146.

Horwitz, D., Mills, D. and Heath, S. (2002) *BSAVA Manual of Canine and Feline Behavioural Medicine.* BSAVA, Quedgeley, UK.

Houpt, K. and Virga, V. (2003) Update on clinical veterinary behaviour. *The Veterinary Clinics of North America* 33, 185–453.

International League Against Epilepsy, Glossary, http://www.ilae-epilepsy.org/Visitors/Centre/ctf/glossary.cfm

Landsberg, G. (2005) Therapeutic agents for the treatment of cognitive dysfunction syndrome in senior dogs. *Progress in Neuro-Psychopharmacology and Biological Psychiatry* 29, 471–479.

Maarschalkerweeerd, J.L., Edenburg, N., Kirpenstein, J. and Knol, B.W. (1997) Influence of orchiectomy on canine behaviour. *The Veterinary Record* 140, 617–619.

Overall, K.J. (1997) *Clinical Behavioral Medicine for Small Animals.* Mosby, St Louis, Missouri.

Overall, K.L. (2000) Natural animal models of human psychiatric conditions: assessment of mechanism and validity. *Progress in Neuro-Psychopharmacology and Biological Psychiatry* 24, 727–776.

Platt, S. and Olby, N. (2004) *BSAVA Manual of Canine and Feline Neurology.* BSAVA, Quedgeley, UK.

Podberscek, A.L. and Serpell, J.A. (1996) The English Cocker Spaniel: preliminary findings on aggressive behaviour. *Applied Animal Behaviour Science* 47, 75–89.

Rijnberk, A. and de Vries, H.W. (1995) *Medical History and Physical Examination in Companion Animals.* Kluwer Academic, The Netherlands.

Rusbridge, C. (2005) Neurological diseases of the Cavalier King Charles Spaniel. *Journal of Small Animal Practice* 46, 265–272.

Studzinski, C.M., Araujo, J.A. and Milgram, N.V. (2005) The canine model of cognitive aging and dementia: pharmacological validity of the model for assessment of human cognitive-enhancing drugs. *Progress in Neuro-Psychopharmacology and Biological Psychiatry* 29, 489–498.

Vaisanen, M., Oksanen, H. and Vainio, O. (2005) Postoperative signs in 96 dogs undergoing soft tissue surgery. *The Veterinary Record* 155, 729–733.

Vite, C.H. (ed.) *Braund's Clinical Neurology in Small Animals: Localization, Diagnosis and Treatment.* International Veterinary Information Service, Ithaca, New York (http://www.ivis.org).

Webb, A.A., Jeffrey, N.D., Olby, N.J. and Muir, G.D. (2004) Behavioural analysis of the efficacy of treatments for injuries to the spinal cord in animals. *The Veterinary Record* 155, 225–230.

Wojciechowska, J.L. and Hewson, C.J. (2005) Quality-of-life assessment in pet dogs. *Journal of American Veterinary Medical Association* 226, 722–728.

Index

abnormal behaviour 231
acoustic communication 219
action potential 91
activity rhythm 156
adaptation 208
aggression 113, 229, 247
aggressiveness 192
ancient dog 44
anxiety 246
appeasement 228
appetite 250
archaeology 22
artificial selection 82
associative learning 122
attention 216–217
auditory cortex 93
auditory system 92

bang phobia 231
BARF 237
barks 113, 115, 217
behaviour 61
behaviour therapy 232
behavioural rating scales 190
behavioural tests 188
biting 247

blindness 253
body size 16
body-language 228
boldness 187
Borophaginae 8
brain 63
brain tumour 252
breed standard 53
breeding pair 153
breeding programmes 168
breeding value 168
breeds 52
breed-typical behaviour 202

Canidae 3
Caninae 4, 10
Canini 11
cardiovascular disease 256
Carnivora 5
castration 236
CBARQ 191
chaining 129
choke chain 238
clade 26
classical conditioning 122
clicker 126

clicker training 235
clinical signs 245
cochlea 93
colour vision 95–96
communication 106
complex traits 79
compulsive disorder 257
control of feral dogs 161
control region 24, 39
cooperative hunting 154
coping style 182, 185–186
cortisol 73
counter-conditioning 141
coyote 107
crossbreeding 34

dangerous behaviour 229
dating of origin 30
deafness 253
defecation 251
denning 158
dental formula 7
desensitization 139
detector dogs 178
development 69
diet 16, 237
dingo 35
disease definitions 244
divergence 44
diversity 38
dizziness 254
DMA 199
DNA 77
DNA sequence 39
dog–human communication 211
domestication 21, 44, 187
dominance 109

ear 92
ear pinch 128
early breeding 48
EBV 168
endocrine disease 254
endocrine management 236
energy 166
epilepsy 247
ethology 61

excitability 194
extinction 134
eye 95

facial expression 114
facial signals 211
factor analysis 210
family groups 108
fatigue 246
FCI 51
fearfulness 190
feeding ecology 157
feelings 183
feral dog 147
feralization 148
fight 113
fitness 170
fixed ratio 130
fly catching 258
forebrain 64
fossil 3
founder 48
fox 11, 87

gastrointestinal disease 255
gazing 215
gene 76
gene expression 81
genome 77
gesture 211, 227
golden jackal 107
group size 154

habitat 161
habituation 121
haplotype 36
harness 239
heredity 81
heritability 82, 178
  in personality 201
Hesperocyoninae 7
hindbrain 64
home-range 155
hormones 68
human bond 159
human-given cues 213

hybrid 82

idiopathic aggression 260
imitation 124
imprinting 92
inheritance 167
instincts 77
instrumental learning 234
interspecific interactions 209

itch 249

jackal 107

key stimuli 66

learned irrelevance 131
learning theory 235
life-span studies 199
limbic system 64
linkage disequilibrium 81
litter size 175

marker 126
Mendelian inheritance 79
meningitis 252
mentality 203
merle 80
MHC 48
microsatellites 40
midbrain 64
mixed-species group 209
modal action pattern 67
molecular clock 45
molecular methods 44
monogamy 108
motivation 71, 125
motivational state 195
mRNA 78
mtDNA 22, 39
mutation 87
muzzle 239

narcolepsy 244, 246
natural selection 166
nature and nurture 62
neocortex 64
neurons 64
niche 207
nociceptor 101
nose 98
nucleotide diversity 29

obedience 235
object name learning 218
olfaction 98
operant conditioning 121
optimum trait value 171
origination event 40
overshadowing 132

pack 105
pain 101, 237, 249
palaeontology 31
parasympathetic neuron 65
parental care 108, 159
perceptional system 70
pessimistic 183
phenotype 79
pheromones 99, 240
phylogenetic tree 41
phylogeny 13
phylogeography 26
play 112
playback 219
playfulness 193
play-sounds 113
pointing 214
population density 151
population genetics 23
positive punishment 137
posture 15
predation 157
prey 157
problem-solving 220
prompting and fading 128
pruritus 249
pseudopregnancy 236
punishment 134

puppy testing 198
QTL 84
quantitative genetics 167

rabies 152
rage syndrome 260
rank order 111
recombination 84
reflex 66
rehome 230
reinforcement 122
reinforcement schedule 129
reproduction 158
resource allocation 169
retina 95
Rico 217
RNA 78
running 15

scavenger 151, 157
seizures 248
selective culling 151
selenocysteine tRNA 15
semi-natural conditions 105
sensitization 121
sequence analysis 22
sex ratio 159
shaping 127
shock collar 239
shyness–boldness 187
side effects of breeding 169
silver fox 87
single gene effect 172
sires 52
skull 5–6
smell, inability 254
sociability 193
social cognition 219
social facilitation 123
social isolation 241
social learning 109, 123
social tolerance 111
social unit 153
socialization 212
solitary dogs 150

stability 184, 197
standardization of tests 203
stereotypies 73
stimuli 66
stimulus equivalence 133
stimulus generalization 132
strange-situation test 210
stray dog 148
stress 72
substitutions 32
sympathetic neuron 65

tail chasing 74, 258
tail docking 102
taste 100
teeth 8
temperament 182
territorial behaviour 156
territory 152
trainability 184, 194
transcriptome 85

ultrasound 94
umami 100
unwanted behaviour 225
urbanization 226
urination 250

variable ratio 130
village dog 150
vision 94
vocal repertoire 117
vocalization 94
vomeronasal organ 99

wall jumping 74
weakness 246
wild type allele 172
wolf 21
word communication 217

Y chromosome 40